Competition and Investment in Telecommunications and Media Markets

Roberto E. Balmer

1

ISBN 978-1495301346

This book is dedicated to my parents, Peter and Mariangela, to my grandfather, Ettore, to Laura, Susanne and Gianna. Without their patience, understanding and support the completion of this work would not have been possible.

The present book builds on a Ph.D. thesis presented on 10 December 2013 to a Doctoral Committee composed of

Prof. Carlo Cambini, University of Lugano

Prof. Emanuele Giovannetti, Anglia Ruskin University of Cambridge

Prof. Antonio Nicita, University of Rome I

Ph.D. tutor:

Prof. Domenico Tosato, University of Rome I

This book reviews the economic literature on cooperative investment in next generation broadband networks and geographic regulation. It additionally proposes innovative models for estimating the level of competition and investment in the fixed telephony market and the retail market for newspapers. In doing so, it addresses two hotly debated issues in business strategy and economic policy: the determinants of investment and competition and the impacts of innovative investment schemes.

The first chapter reviews the literature on new cooperative investment schemes in next generation broadband networks and geographic regulation. The effects on competition, investment and welfare of such schemes crucially depend on the details of the agreements. For instance, in the case of joint-ventures, the manner in which investment costs are shared and internal and external access prices are determined significantly impacts the outcome. In the case of long-term access agreements, it is essential to consider how access tariffs are structured, whether they can adapt to market developments ex-post, and whether contracts are signed before or after the investment takes place. Generally, many of these agreements allow for some extent of risk sharing, offering the possibility of increasing investment incentives when firms are not risk neutral. It is suggested that regulators consider introducing regulated co-investment agreements complementing current regulation, in addition to considering geographically segmented access prices.

The second chapter assesses entry and competition in local retail markets for newspapers. It builds on the new empirical industrial organisation (NEIO) literature to estimate sustainable coverage and competitive effects of entry for Swiss newspaper sellers which sell composite goods (newspapers, food and other goods of daily use). An entry threshold ratio methodology is used, allowing for model estimation even when the range of products under examination is not exactly defined and when price and quantity data are not available. It is found that under duopoly prices the market size of a Commune required for single firm entry is about twice as large as under monopoly prices. A clear and quantifiable trade-off between competition and investment therefore exists. Moreover, it is found that while a second entrant in this market strongly increases competition, further entry doesn't have a significant additional competitive effect. From a welfare perspective, therefore, it can be stated that "two is enough" to ensure competition in this market.

In the third chapter, competition and market strategies in the Swiss fixed telephony market are assessed. A market model based on a generalised version of the traditional "dominant firm – competitive fringe" model, is developed. Direct estimation of the incumbent's intertemporal residual demand function is performed by instrumenting the market price with incumbent-specific cost shifting variables, as well as other variables. The concrete estimates show that residual retail demand for voice traffic is highly inelastic. Such a level of elasticity is only compatible with a profit maximising incumbent in the case of largely competitive conduct. It is therefore found that the Swiss incumbent acted largely competitively, and that current regulated telephony retail price caps could not be justified on the basis of a lack of competition.

Chapter I

Geographic regulation and cooperative investment in next generation broadband networks

A review of recent literature and practical cases

Alternative telecom operators have continuously invested in their own infrastructure in recent years. After more than a decade since liberalisation, competitive conditions have substantially changed, especially in urban areas. European regulatory authorities have acknowledged this development by starting regional deregulation. Additionally, different forms of cooperative investments in next generation broadband have appeared on the market. The effects of such schemes on competition, investment and welfare crucially depend on the fine details of implementation. For instance, in the case of joint-ventures, it matters how investment costs are shared and how internal and external access prices are determined. In the case of long-term access agreements, it is essential to consider how access tariffs are structured, whether they can adapt to market developments ex-post and whether contracts are signed before or after the investment takes place. Generally, many of these agreements allow some extent of risk sharing, offering the possibility to increase investment incentives when firms are not risk neutral. This chapter reviews the theoretical and empirical literature on geographic regulation and co-investments in next generation broadband. It is suggested that regulators consider introducing regulated co-investment agreements complementing current regulation or in some cases even substituting for it, in addition to considering geographically segmented access prices.

1.Introduction

The continuous investment of alternative operators in telecommunications infrastructure in the years after liberalisation has led to increasingly differing competitive conditions across geographic areas. This is particularly the case in those network segments where alternative operators have invested; in national and regional backbone segments and also increasingly in local access directly connecting households in urban areas with next generation broadband. The latter investment may be seen as particularly valuable as high speed broadband has substantial positive spill-overs for the economy (Bourreau, Cambini and Hoernig (2012a) review relevant literature and estimates). Given that the regulators' main objective is to ensure competition, uncertainty arises about whether a nationally uniform regulatory approach remains valid or whether some form of regional deregulation would be warranted. Positive spill-overs from investment for the economy may reinforce this uncertainty. In Europe deregulation in dense, more competitive areas has accordingly increasingly been undertaken. The regulatory options a regulator has to implement this may range from regional full deregulation to access only obligations or forms of price regulation and will be reviewed in chapter 2 as well as their effects on competition, investment and welfare.

In addition, firms as well as regulators seem to start to understand that network duplication, which traditional infrastructure competition has sometimes implied, is inefficient from a welfare point of view as investment costs are also duplicated. A natural solution is the use of cooperative investments whereby an infrastructure able to host both partners is rolled-out. Such co-investment schemes may also be used to distribute and share investment risk between the partners implying higher investment incentives, leading to higher quality broadband and more innovation. The presence of such co-investment agreements increases the complexity of the assessment of competition and investment incentives substantially, as the details of such agreements matter. In particular, allowing some co-investment clauses may be welfare optimal, while others may restrict competition too strongly (e.g. an high internal or external access price). Chapter 3 reviews the literature on cooperative investment in next generation broadband, considering the fine details of these mechanisms, as well as possible regulatory options such as the introduction of regulated joint-ventures in which the firm rolling out must offer the entrant the option to join it in a joint-venture at equal conditions[1]. The development of the literature on these topics is still a work in progress, as the introduction of regional regulation took place only around 2008 and large scale broadband co-investment agreements began only around 2009 – less than half a decade before this paper was written. Given the complexity of such agreements, many questions still remain open.

Both geographic regulation as well as co-investments take place in a context of migration from legacy to next generation access (NGA) networks[2]. Traditional copper networks will be only progressively substituted by next generation infrastructure, and the regulation of both legacy and next generation infrastructure may affect this process and, in particular, investment incentives. Bourreau, Cambini and Hoernig (2012a) review the literature on migration. Most importantly, Bourreau, Cambini and Doğan (2012) find that regulated legacy access charges may affect investment in NGA in different ways. While an increase in the regulated access price to the new network in all cases increases investments, the effects of access prices associated with the legacy network are unclear. The authors show that with a high legacy network access charge:

---

[1] A generic overview on the effects of access regulation on investment incentives is provided by Cambini and Jiang (2009). Also, a high-level review of the literature on geographic regulation, co-investments and migration may be found in Bourreau, Cambini and Hoernig (2012a).

[2] Also NGN, considering any type of next generation network not only related to access networks

i)   the entrants' opportunity cost of investment is low, increasing its investment incentives (replacement effect)

ii)   the incumbent risks to lose (or cannibalize) wholesale profits (wholesale revenue effect) from an investment (it is assumed that an entrant can more easily roll-out its own network infrastructure once the incumbent has deployed it (investment spill-over).

iii)   pressure on retail prices for legacy network based services is low. When the access price is low instead, as long as next generation services are seen as substitutes, the overall profitability of the investment is reduced (business migration effect).

Overall, it is therefore unclear whether a relatively high legacy network access charge can increase investments in next generation broadband or not[3]. A high legacy access charge increases investment incentives of the entrant and sometimes those of the incumbent, potentially increasing dynamic efficiency, while negatively affecting static efficiency. The welfare maximising access prices a regulator should set in case of regulation of the legacy network are then shown to depend on the market environment and in particular on the amount of investment spill-overs (with high spill-overs the regulator would set a high access charge to counterbalance the negative effect it has on investments of the incumbent). Finally, when setting both copper and fibre access prices, these effects interact. Whenever a legacy network is present in the models reviewed, such migration issues are considered in some way. Most papers that will be analysed in this survey assume, however, given regulated marginal cost access to the copper network for all operators, implying absence of rent from this infrastructure minimizing distortions.

This paper consists of two major sections that explore different theoretical issues related to the deployment and regulation of next generation broadband networks in Europe. Chapter 2 introduces geographic segmentation of regulation, reviews regulatory principles and practices in Europe as well as the theoretical and empirical literature on the subject. Chapter 3 describes different types of co-investment agreements for the roll-out of next generation broadband networks in Europe and describes regulatory principles and practice. In addition, theoretical and empirical literature on the subject will be reviewed. Chapter 4 concludes the paper, and integrates ideas in the two prior chapters.

2. Geographic segmentation of regulation

The cost of rolling-out fixed access infrastructure is typically related to population density which in turn varies strongly across areas. Such geographic differences in investment costs may lead to geographically different market structures such as a higher number of entrants in urban areas. Increasingly competitive conditions in different geographic areas start to differ within European countries. As effective competition is the main objective of telecoms regulation, there is an ongoing debate about whether full or partial deregulation of geographic areas under increased competition is socially optimal. Since the liberalization of the telecoms market alternative operators are investing in their own network infrastructure. This is especially the case with the roll-out of NGA infrastructure, as explained in the European Commission Recommendation on regulated access to next generation access networks[4]. Consequently, the coverage

---

[3] Hoernig, Jay, Neu, Neumann, Plückebaum and Vogelsang (2012), a report for the European Competitive Telecommunication Association (ECTA), states that a high copper access charge reduces investment incentives focussing on the wholesale revenue effect. Plum (2011), a report for the European (incumbent) Telecommunications Network Operators (ETNO), instead states the contrary focussing on the business migration effect.

[4] The 2010 NGA recommendation states for these cases that "where the incumbent deploys FTTH, NRAs should in principle mandate unbundled access to the fibre loop. Any exception could be justified only in geographic areas where the presence of several alternative infrastructures, such as FTTH networks and/or cable, in combination with competitive access offers is likely to result in effective competition on the downstream level".

of regional alternative networks as well as their number has increased over time. Authors such as Cave (2008) argue that this must trigger a geographically differentiated regulation[5].

While it is always difficult to draw direct inferences on the effects of regulation from the market outcome, it is convenient to describe some fundamental market characteristics at this stage.

Download speeds via the legacy network (xDSL[6]) vary significantly across Europe (**Figure 1**). While the average xDSL speed in 2012 was 7.23 Mbps (around 35 Mbps for Cable and 37 Mbps for fibre to the home (FTTH)), speeds in Denmark were on average 11 Mbps while those in the Slovak Republic were 3 Mbps[7]. The major cause of slow DSL speeds is insufficiently upgraded backhaul networks. While on aggregate Europe scores well when compared to the US[8], other sources show that comparisons with countries such as South Korea or Japan are less favourable[9]. While this may also be a consequence of different population densities or customer preferences it can also be the result of lacking investment incentives in higher speed access networks generated by access regulation and in particular also by geographic regulation (and co-investments) or its absence.

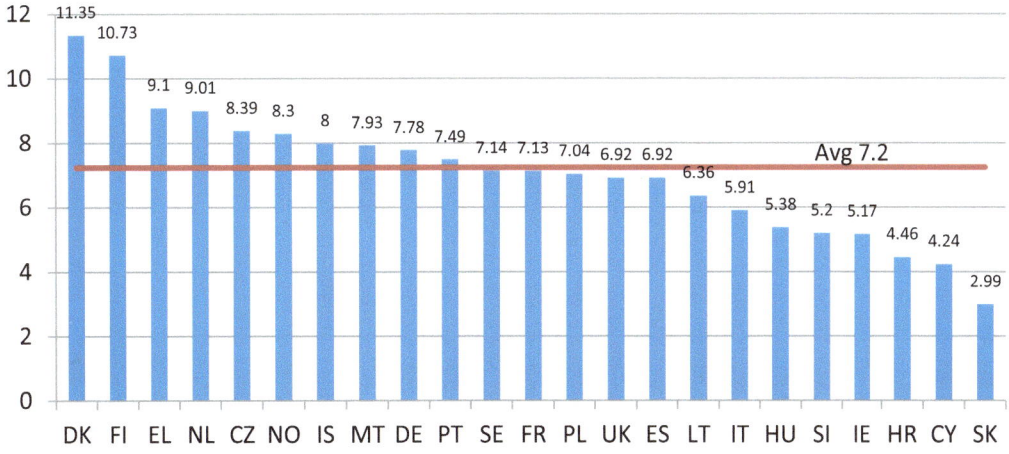

Figure 1 – Actual xDSL Speeds in Europe (Source: Samknows, March 2012)

We will see that pioneering NRAs in this field include Austria, Portugal and the UK. From the aggregate data these countries do not seem to have a particularly high or low performing broadband infrastructure when compared to other European countries. It should, however, be noted that such geographic deregulation efforts are relatively recent and concerning only strongly limited areas. Any impact on infrastructure investment at national level may therefore still be limited.

It may be interesting to point out that overall broadband access prices do not seem to be higher in countries with higher xDSL performance on the market. Van Dijk (2012) shows for instance that at speeds between 12 and 30 Mbps prices in Italy and Ireland, where few infrastructure investments in xDSL seem to have taken place are also higher (around 43€ and 45€ per month[10]) than in Denmark and Finland (29 and 35€ per month) for the median offer. This also holds when comparing the least expensive offers in

---

[5] In particular Cave proposes to distinguish three areas ("potentially competitive", "probably monopolistic but where NGA investment can be commercially justified" and "non commercial") regulated by principles of "forebearance", "mandatory access to dominant NGA" and "mandatory access to one or more collectively dominant NGAs" respectively.

[6] xDSL describes all digital subscriber line based technologies such as ISDN, ADSL or VDSL. These are copper-based.

[7] Industry average speeds are not calculated for Europe

[8] Actual download speeds in the U.S. are 5.3 Mbps for xDSL, 17 Mbps for Cable and 30 Mbps for FTTH

[9] Akamai (2012)

[10] In €/PPP (VAT incl.), see p.116 and p.84

Italy in Ireland (around 26€ and 29€ per month) with Denmark and Finland (around 24€ and 25€ per month). The same is true for lower speeds at 2-4 Mbps[11]. When comparing national population densities the picture is not coherent. For instance Finland has a very low population density (44 per sq mi) and Italy a very high density (512 per sq mi), while Denmark and Ireland have an intermediate density (333 and 153 per sq mi respectively). This suggests that it may be insufficient to compare nationally aggregate market outcomes. For instance population density in Helsinki is not lower than in other capitals. Until recently only few disaggregated data was available. The increasing adaption of regulation to geographic market conditions and the will to support investments locally has, however, led to a recent increase of monitoring. The European Commission has asked Point Topic to map progress with next generation investments in members states and regions (30 mbps or above). **Figure 2** shows NGA coverage in urban and rural areas. Overall NGA coverage seems to be highest in relatively dense countries such as the Netherlands, Switzerland and Belgium. At the same time these countries have historically strong cable competitors. In addition NGA coverage in rural areas is in all (large) countries significantly lower. This digital divide seems, however, still to be stronger in countries without a historical cable competitor.

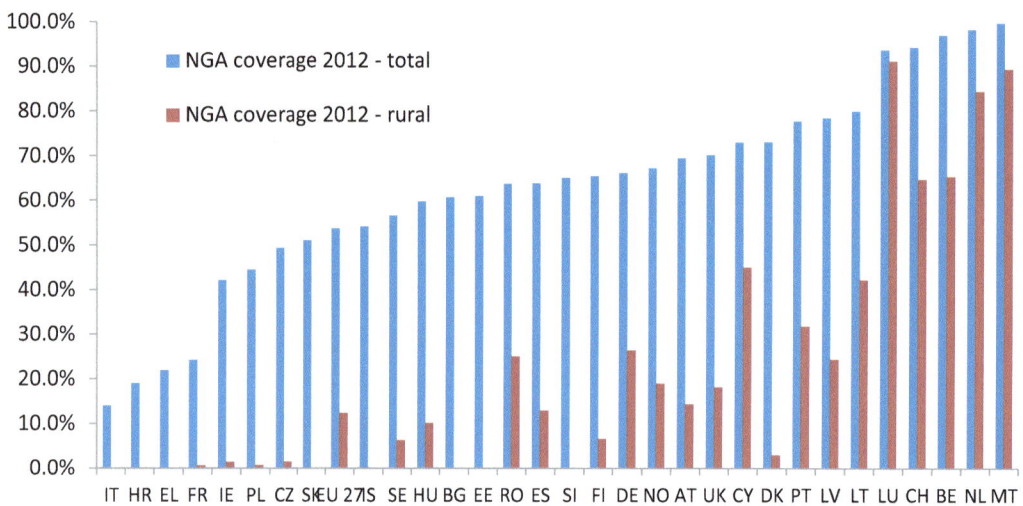

Figure 2 – NGA coverage in Europe, total and rural areas (Source: Point Topic)

The following sections will describe the regulatory principles at EU level guiding regulatory action in this field and regulatory practices implementing geographic segmentation of regulation. In a subsequent section the academic literature is reviewed.

2.1.   Regulatory principles in the European Union

Geographical market analysis has always been a part of the European regulatory framework. It states that even if demand and supply-side substitution patterns may suggest a national market, sub-national markets can be defined when competitive conditions differ to a sufficient extent (e.g. urban and rural)[12].

---

[11] There is no data in this category for Italy though.

[12] European Commission (2002), 56.: *"the relevant geographic market comprises an area in which the undertakings concerned are involved in the supply and demand of the relevant products or services, in which area the conditions of competition are similar or sufficiently homogeneous and which can be distinguished from neighbouring areas in which the prevailing conditions of competition are appreciably different. The definition of the geographic market does not require the conditions of competition between traders or providers of services to be perfectly homogeneous. It is*

This approach will be referred to as geographic segmentation of markets. In this case it is possible that the absence of significant market power of a firm or firms in a geographic sub-market can be demonstrated. In such cases, the regional market would then not be subject to any kind of asymmetric regulation anymore (full deregulation). Moreover, lighter remedies can be imposed in sub-areas with stronger competitive constraints within an area where significant market power is found. This approach will be referred to generically as geographic segmentation of remedies. While it will be shown that the high flexibility with remedies means that technically this difference may not be of fundamental importance, the regulatory processes which lead to one or the other are – in Europe – fundamentally different. Finally, the aggregate of both approaches will be referred to as geographic segmentation of regulation or geographic regulation.

A series of national regulatory interventions regarding geographic segmentation of markets and remedies have been notified to the European Commission since 2008. While the European Commission currently has veto power on member states decisions on market analysis issues (i.e. in this context the definition of geographic markets), this is to date not the case on remedies (i.e. in this context the geographical differentiation of remedies).

The BEREC Draft Common Position on Geographic Aspects of Market Analysis[13] (2013) acknowledges the described market developments and sees an increasingly importance of geographically differentiated regulation in Europe. BEREC (2013) aims at giving European NRAs guidance on geographic regulation[14] and follows an earlier Common Position of 2008.

The Common Position states that NRAs should consider making a detailed geographical market analysis when some key indicators are present:

- *One or several alternative operators have significant but less than national coverage and exert a significant competitive constraint at the retail level in the areas where they are present*

- *The incumbent operator differentiates retail prices geographically or the incumbent operator is setting a national uniform retail price but there are significant price differences between the incumbent operator and alternative operators where the latter is present; and*

- *There are significant geographic differences in product characteristics*

The telecommunications sector consists of complex markets and technical products. For a detailed description of the markets and products analysed in the upcoming sections the reader may refer to BEREC (2010a). In recent years operators have been increasingly climbing the ladder of investment being able to replicate for example wholesale broadband access (WBA) products based on local loop unbundling (LLU). Also, in several countries independent alternative operator technologies (cable, FTTx, mobile broadband) are expanding rapidly allowing the provision of similar services. Provided that the described technologies are found to be retail product substitutes, indicators have to be analysed hinting to regionally different competitive wholesale conditions. The Common Position states that most likely candidates for segmentation are the WBA and leased line markets (wholesale services). Another likely candidate would be the market for physical access to the end customer[15] (essentially LLU).

---

sufficient that they are similar or sufficiently homogeneous, and accordingly, only those areas in which the conditions of competition are 'heterogeneous' may not be considered to constitute a uniform market."
[13] BEREC (2013), public consultation version
[14] A similar document has been product at OECD level (OECD, 2012)
[15] Under the so called "modified Greenfield approach" regulation on the market under examination is disregarded, but regulation on other (upstream) markets is treated as exogenous. I.e. an analysis of the competitiveness of the WBA market will consider LLU regulation to remain in place.

The Common Position distinguishes two types of countries. First there are countries - especially in Western Europe - where competition was mainly driven by LLU-based market entry and only partially by alternative infrastructures such a as cable (**scenario** 1). Secondly, there are countries - especially in Eastern Europe - where it is mainly driven by alternative infrastructures such as cable (**scenario** 2). Romania is an interesting example, as in the broadband retail market intense competition of the incumbent with cable operators is taking place. Today cable operators hold a higher share of the retail market than the incumbent. In addition there are regions where even two cable operators are present. The reason for this situation may be that the incumbent was slow to enter the broadband market and when it did, it did not enter aggressively (also because it has to offer uniform retail prices while cable operators are only present in urban areas). An additional reason may be that regulation on the incumbent was introduced only recently meaning that DSL-based competition was less aggressive than in other countries[16].

In case a geographical segmentation of the market is indicated the Common Position suggests choosing adequate geographic units. Generally there are two approaches: political/administrative boundaries or a network approach based on the topology of the incumbent operator. In any case, the Common Position states that the units should satisfy the following four conditions

> a. *Mutually exclusive and less than national*
> b. *The network structure of all relevant operators and the services sold on the market can be mapped onto the geographic units*
> c. *Have clear and stable boundaries*
> d. *Small enough that the competitive conditions are unlikely to vary significantly within the unit but at the same time large enough that the burden on operators and NRAs with regard to data delivery and analysis is reasonable[17].*

Homogeneously competitive areas should then be aggregated from the chosen geographic units. Homogeneity is judged essentially with the following criteria:

> a. *The barriers to entry in the market*
> b. *The number of operators that exert a relevant competitive constraint on the SMP operator*
> c. *The market shares of the SMP operator and the alternative operators*
> d. *The prices*

Typically geographic areas in scenario 1 could, for instance, be defined as the areas covered by unbundled MDFs[18] (e.g. UK WBA case UK/2010/1123 in Table 1). Depending on the extent of alternative parallel networks the segmentation could also be made based on the alternative networks topology or administrative geographic areas. In scenario 2, such areas could be based on administrative geographic areas (for example communes (e.g. Polish WBA case PL/2011/1184 in Table 1) or municipalities (e.g. Czech WBA case CZ/2012/1322). In addition, where a vertically-integrated cable operator is present, the

---

[16] Informa (2011)
[17] As noted in the Common Position, if the choice of a geographic unit that is too small may lead to a very significant number of units (even in the thousands). While the aggregation of geographic areas may contribute to solve part of the administrative burden derived from this fact, it is nevertheless a factor that may have to be weighted carefully by the NRA before deciding on the appropriate geographic unit.
[18] Main distribution frames, in practice at the local exchange facility.

competitive effects on the wholesale market need to be considered only to the extent that they are relevant[19].

In practical cases, regulators often analysed whether the incumbent operator in a given area would have a market share below a certain threshold (e.g. 40-50%) and whether sufficient alternative infrastructures existed (number of players). More concretely the BEREC report on co-investment and significant market power (SMP) in NGA networks notes that "a market characterised by two operators implies automatically that one of the players disposes of a market share of 50% or more and that it is therefore to be expected that a market with high entry barriers with one or two operators in the market raises concerns about dominance and more generally the competitive situation of the market." Conversely, it is concluded that only markets with three or more independent operators can lead to effective competition in the physical access market in such an environment. An access market consisting of two infrastructures (e.g. incumbent and cable) is therefore generally not being considered to be sufficiently competitive. There are, however, various cases in Switzerland and France (usually as a consequence of co-investment) where three or more independent infrastructures currently co-exist (e.g. Basel, Paris, Zurich). In a full market analysis assessing the level of competition several other important factors next to the number of players, such as entry barriers, market shares, downstream competition, indirect effects and commercial or regulated wholesale products (e.g. often not given for Cable) would need to be assessed, however.

As has been shown sub-national geographical markets are defined in case they are indicated by demand and supply side substitutability analysis or in case of sufficiently heterogeneous competitive conditions. The resulting sub-national markets must in turn be sufficiently homogeneous and have stable borders themselves. Typically, the Common Position states, geographic market segmentation is applied when an national regulatory authority (NRA) believes that some (non SMP) areas are competitive enough to fully withdraw regulation. Finally, it should be noted that possible closing of redundant traditional local exchanges (MDF) during the migration to NGA network may have consequences on the geographical market definitions.

In case that the heterogeneity of economic conditions is not sufficiently strong to justify geographic markets or where the borders of the market are not sufficiently stable or sustainable the Common Position suggests the definition of a national market with the imposition of − more flexible - geographically differentiated remedies. In these cases, typically no fully deregulated areas are defined. Interestingly, the full or partial deregulation of an area may according to the Common Position also have an economic impact in the remaining areas in case of cost-based regulation. In case of a segmentation of markets, it is likely that deregulation could take place in dense, low cost areas leaving only the higher cost areas subject to regulation, featuring a network with a higher cost base per user and higher regulated average prices than before the deregulation of urban areas[20].

2.2. Regulatory practices in Europe

Since the first decisions imposing geographic segmentation of regulation (UK and Austria in 2008) a number of European national regulatory decisions in this field have been added (Table 1). For a more detailed review of WBA geographic segmentation of regulation in Europe the reader can refer to Houpis et al. (2011). For a review of the approach to geographic segmentation of regulation in the U.S. the reader can refer to Stockdale (2011). Finally, for a review of worldwide cases covering also countries such as Australia the reader may refer to Xavier & Ypsilanti (2011).

---

[19] Whether technologies are retail substitutes and whether they can indirectly constrain the wholesale market under consideration in case no wholesale product is offered (e.g. Cable) needs to be analysed in detail. See also BEREC report on self-supply, BEREC (2010b).
[20] See also BEREC Common Position 2013

This section will review recent decisions and summarize the current situation of geographic regulation in Europe across all communications markets. In particular, proposed and implemented geographic access regulations in European member states in the following markets will be analysed: i) wholesale broadband access, ii) wholesale leased lines and iii) wholesale (physical) network infrastructure access. Detailed references to the regulatory decisions summarized below can be found in Table 5.

In a first decision proposal the **Austrian NRA** originally wanted to introduce geographically segmented markets as the first NRA in 2008 in the WBA market. The European Commission (EC) had signalled to veto this decision as the boundaries of this market seemed unstable. The NRA had then adapted its proposal to define a national WBA market and proposed to withdraw most remedies in the more competitive segments of the market. Lighter remedies were proposed to be imposed in MDF areas with two or more alternative operators present, incumbent market share below 50% and serving more than 2'500 households. The European Commission had accepted this proposal[21]. Regarding remedies it stated that "*the geographic differentiation of remedies may be appropriate in those situations where, for example, the boundary between areas where there are different competitive pressures is variable and likely to change over time, or where significant differences in competitive conditions are observed but the evidence may not be such as to justify the definition of sub-national markets*". The imposition of geographical remedies was then, however, rejected by the Austrian Administrative Court on 12 August 2008 leading to an implementation of regulation without geographical differentiation (without lighter remedies in more competitive areas). In the recent fourth round of market analysis (2013) RTR again proposes a national market, this time with uniform remedies (retail minus price control; products are restricted to only business-grade products). The proposal is still pending.

On the market for leased lines instead, the **Austrian NRA** proposed in 2008 a geographic segmentation of markets of high speed (>2 mbps) terminating segments of leased lines in two geographic markets: 12 competitive cities and the rest of the country. The cities would be those communes having i) population of more than 15'000, ii) more than three operators offering terminating leased line segments based on own infrastructure and iii) a market share of the incumbent <50%. The European Commission stated, however, that it would have doubts about the homogeneity of competitive conditions within these markets and that the incumbent could well not have SMP also in the rest of the country for high speed leased lines. In particular, more information about the geographical distribution of market shares and pricing structures as well as their evolution over time was been requested. The European Commission also reminded that a defined market should have stable boundaries over time. The decision has then been withdrawn implying that currently high speed wholesale leased lines are deregulated in 12 cities but not (yet) in the rest of the country. In its more recent fourth round market analysis (2013), the Austrian NRA reverted back to a national market and uniform remedies. The European Commission vetoed this decision as there seems to be a lack of evidence for homogeneous competitive conditions across all regions in the country. BEREC has shared this view. Especially from the 2008 analysis in the mentioned 12 cities TAs market shares are low between 23 and 34% in the relevant urban market, while the incumbent would not face significant competition in more rural markets. The European Commission has asked for an updated and a detailed analysis. Also the European Commission argues any reregulation should be carefully evaluated.

The **Czech NRA** in 2012 proposed for the WBA market two geographic submarkets: districts where at least three infrastructures are present and the incumbent has less than 40% market share and other districts. Consequently, it proposed to fully deregulate the area under infrastructure competition while

---

[21] "*Based on the general principle that remedies should be tailored and proportionate to the identified competition problem, it can be appropriate for NRAs to impose remedies which take account of locally/regionally differentiated competitive conditions while retaining a national geographic market definition.*"

continuing to regulate the rest of the country with relatively light remedies excluding cost-orientation. The European Commission stated that this proposal is mainly based on the number of independent networks and therefore insufficient. It stated that for instance the incumbent's wholesale offer would be national with national prices. Also, the homogeneity of competitive conditions would seem not to be given within the "urban" areas as they seem to include also some small cities (with lower economies of scale). Moreover, the European Commission had doubts about the competitiveness of such areas. In particular it doubted whether Wi-Fi networks may be retail substitutes to DSL as Wi-Fi coverage would be limited and offer only lower speeds. It also stated that indirect constraints on the wholesale market would be unlikely to be sufficient for Wi-Fi as well as for Cable. While BEREC had supported the NRAs proposal, it was vetoed by the EC.

In 2009 the **Dutch NRA** has formally notified a national market including copper and fibre local loops and national remedies were set. Binding price caps for fibre, however, were in practice set per cost area (NL/2009/0868, NL/2013/1439). For unbundled optical distribution frame access to FTTH lines and ancillary services such as backhaul and collocation the NRA proposed to take as a starting point the concrete FTTH business case of Reggefiber, the joint-venture formed by the incumbent and an alternative utility operator, to roll out the NGA network (including an effective and not hypothetically efficient capital expenditure as in LRIC). The authority has decided to allow the joint venture to generate a reasonable rate of return including a risk premium. The fundamental idea is to set a first year access price such to make the investment viable (profitable) in a discounted cash flow (DCF) model estimating cash inflows (the revenues of an FTTH model over the assumed lifetime of the network) and cash outflows (capital expenditure and operational expenditure). Assuming that (real) access prices remain constant over the lifetime of the investment, the initial regulated price cap for access products is calculated such that the net present value of future cash flows is equal to the initial capital investment, when applying an initial (reasonable) rate of return (between 7-10%, the exact initial amount is not disclosed). Over time the market environment may then change, e.g. demand, costs and competition may develop positively or negatively for the operator and the internal rate of return (IRR) may then varies over the years. However, such profits are not to exceed the standard risk cost of capital (WACC), increased by a risk premium for fibre, by more than 3.5% (representing regulatory risk). As long as this is not the case (verification every three years) maximum access prices are allowed to remain constant in real terms, i.e. to increase over time along with the consumer price index (1.5% per year). If instead the IRR is too high the prices are adjusted downwards by the authority[22].

The main inputs into this cost model include the expected economic lifetime (25 years), the expected penetration rate (60% after 2 years), capital expenditure per area, the operating costs (12-18€ per line per year), revenues and an initial reasonable rate of return (7-10%). In case of too pessimistic expectations (of demand for instance) the price would be set such (high) that too high profits could be generated. In the converse - too optimistic - case instead profits would ex-post be too low and investment incentives would be adversely affected. Using the DCF model the NRA can adjust its prices over time when expectations turn out to be wrong. While this is a highly flexible setting, targeting essentially regulatory, cost and demand uncertainty of investment over time and flexibility regarding the offering of different price schemes (e.g. volume discounts), it was also decided to set geographically different price caps for 14 areas with differing average capital expenditure requirements. Across these cost clusters fibre unbundling prices 2013 vary substantially from 15.52€ to 25.99€ per line per month in 2013 (Autoriteit Consument en Markt (2013)[23]). In addition to these tariffs, however, there is a national tariff scheme (18.84€ per line per month), which is calculated as the weighted average of all areas. Wholesale customers can choose between the national tariff scheme or the local tariff scheme, but the choice cannot vary from area to area.

---

[22] The approach is broadly described in OPTA (2008)
[23] See Bijlage B

However, in the longer term this may imply that firms are active either in urban areas where they choose the local tariff (as it is lower than the national tariff) or in rural areas, where they choose the national tariff (as it is lower than the local tariff). Interestingly in the long term the binding prices in rural areas could then be lower than the price necessary to cover all area costs as calculated by the business case. In line with what will be show in the next chapter cost recovery prices are reasonably reduced in urban areas, while, however, not increasing them in rural areas.

It should be noted here that the DCF results could also be largely achieved with traditional LRIC pricing as long as identical information is used[24]. Both approaches consider initial capital expenditure, forecast demand developments and use a WACC to calculate the revenues/prices for the first year. The European Commission has in any case accepted the Dutch regulation proposal. To date the Dutch NRA is the only NRA applying geographically differentiated regulated prices according to cost clusters. Interestingly up to now the incumbent had consistently priced about 2.50€ below the price cap, which seems to indicate the presence of relevant infrastructure competition with cable and regulated copper products (Middleton and Van Gorp (2010)). Next to the European Commission notifications the reader may refer to Middleton and Van Gorp (2010) for a review of the Dutch case.

The **Finnish NRA** is facing a particular market with a large number of regional incumbents. Initially, geographic markets were defined along traditional operating areas, where 27 regional incumbents have a market share of more than 90% in their wholesale physical network infrastructure access markets. After the NRA had identified that the regional incumbents started to invest in fibre networks outside of their traditional operating areas, a more refined concept of regional markets was defined. The NRA has started by analysing the competitive conditions in 336 municipalities. It has then aggregated these municipalities based on the following criteria: i) the municipalities compose a physically contiguous geographic market area; ii) in terms of the number of local loops, the market share of the area's market leader in the municipalities belonging to one area is more or less equal (variation of ± 10%); and iii) the number of competing telecommunications operators owning their own local loops in municipalities belonging to the area is more or less equal (± 1 telecommunications operator). The result was the definition of 111 sub-national markets for both the WBA market and the wholesale (physical) network infrastructure access market. In 2012 the NRA in a corresponding full market analysis found seven of these WBA markets to be effectively competitive and full deregulation in these areas was proposed (including Helsinki). In the remaining 104 areas light regulation excluding cost-orientation for WBA of the regional 27 incumbents is proposed. The European Commission did not comment on these issues and the decision has been adopted.

In the market for wholesale (physical) network infrastructure access instead the **Finnish NRA** has defined 111 sub-national markets as in the analysis of the WBA market. None of these markets was, however, deemed to be sufficiently competitive and regional incumbents (at least larger ones) are subject to cost-based regulation. The European Commission did not comment on these issues and the decision has been adopted.

The **German NRA** in 2010 has analysed the WBA market and had identified 771 MDF areas (covering about a quarter of all households) where i) the incumbent has less than <50% retail market share, ii) there are at least four operators offering DSL and iii) the MDF has more than 4'000 subscribers (i.e. sufficiently large to allow unbundling to efficient entrants). However, while the UK and Portugal had proceeded with full deregulation in similar areas the German NRA did not follow this approach and propose a national market. The reasons include that the incumbent pursued a national pricing and product strategy. While the NRA did not propose a geographically segmented remedies it proposed uniform light set of remedies at

---

[24] See Neu, Neumann and Vogelsang (2012), p. 69

national scale, i.e. excluding cost-orientation. The European Commission agreed that there is no conclusive evidence for a geographically differentiated regulation. The decision has been adopted.

The **Italian NRA** has analysed the competitive conditions in the WBA access market in 2011 and concluded that these are not sufficiently heterogeneous to warrant the definition of sub-national markets. The NRA has, however, proposed to differentiate remedies between areas with infrastructure competition and areas without (details to be defined in a later decision). The Commission advised the NRA to follow the criteria for NGA remedies in the NGA recommendation. It reminded the NRA that for a definition of geographic markets the number of operators in a given exchange area, the size of the area to ensure possible entry at the given scale, the distribution of market shares and geographic pricing would need to be analysed. A separate proceeding on geographically differentiated remedies will be opened. With regards to remedies the NRA plans to impose a lighter form of price control in more competitive areas leading to higher prices.

The **Polish NRA** in 2012 has proposed a national WBA market with a lighter set of regulatory remedies for the four largest cities (where there is retail competition from cable and other operators) and one for the rest of the country. In the first segment of the market, mostly urban areas, the NRA proposed to remove the remedies of cost-orientation, accounting separation and transparency, leaving only access and non-discrimination obligations. The European Commission has recommended that the NRA withdraw the proposal and strengthen its analysis of competitive conditions. While the European Commission has no veto on remedies, the NRA has withdrawn its decision proposal.

Also in the WBA market but in an earlier round of analysis of the WBA market of the **Polish NRA** than the one cited above geographically segmented markets with a fully deregulated area of 11 cities under competition. The European Commission had, however, vetoed this proposal as differentiated prices and market shares as well as indirect constraints and potential competition would not have been sufficiently demonstrated and market data had been judged to be outdated.

The **Portuguese NRA** suggested in 2010 a WBA geographic market definition with competitive MDF areas where there is at least one ULL based alternative operator and a cable operator (taken into consideration when the percentage of connected cable households is at least 60% in the area) and non-competitive MDF areas on the other side. The competitive area was proposed to be fully deregulated. The non-competitive areas would still feature a form of light price regulation (retail-minus approach). The European Commission had raised concerns that in some competitive MDF areas the market share of the incumbent is still above 50%. It has therefore invited the NRA to carefully monitor the evolution of competitive conditions in the future, but the decision was not vetoed and has been adopted.

In the leased line market, instead, the **Portuguese NRA** proposed in 2010 a geographic segmentation of the trunk segments of leased lines market (which usually connect the exchanges of the country) in a competitive trunk market connecting 110 local exchanges where at least two alternative operators are present with own infrastructure and one "non competitive" trunk market connecting the rest of the exchanges. It was then proposed to fully deregulate the competitive leased lines routes (as done by the Swiss NRA) and to impose regulation including cost orientation on the remaining lines. Given that the terminating segments are regulated similarly, the regulatory outcome would be similar to the one in Switzerland. The European Commission has stated that the geographical market segmentation is primarily based on the number of operators, which it considers to be insufficient, and that further evidence is necessary, such as markets shares over time and regionally differentiated wholesale and retail pricing. Given the important differences in market shares and network duplication the Commission did however not contest the decision and it was adopted. It invited the NRA, however, to base its next market analysis on more detailed data.

The **Spanish NRA** had identified in 2008 differing competitive conditions in the WBA market but these were not deemed sufficient for a definition of regional markets. It argued that the incumbent's retail pricing was still national. It was also argued that the current NGA roll-out would affect the boundaries of possible geographic markets meaning that sub-national market boundaries would be unstable. However, unlike the German NRA, the Spanish NRA proposed geographically differentiated remedies. In areas where the incumbent faces infrastructure-based competition (at least Cable and at least two LLU-based competitors) and where the incumbent's market share is below 50%, the NRA proposed the withdrawal of the cost-orientation obligation. The Commission has asked the Spanish NRA to detail its geographic analysis further by analyzing different geographic commercial strategies, average retail prices, functionalities provided and market shares (retail and wholesale) in both areas. Also the Spanish NRA was suggested to analyse in detail the stability of boundaries and a possible trend to competition in the urban areas. BEREC supported the Spanish NRAs view of a national market, in particular because of the unstable character of geographic borders and the fact that different retail prices could reflect different technologies rather than market pressure[25]. It also agreed that competitive differences could warrant geographic differentiation of remedies. Finally, however, the Spanish NRA has withdrawn the proposal imposing remedies formerly proposed only in more rural regions also on national scale. No further round of market analysis has yet been notified.

In the leased line market the **Swiss NRA** in 2010 had defined the market for trunk segments of leased lines as the market of lines between Communes where two or more alternative operators to the incumbent are present with own infrastructure (e.g. 25 Communes in 2009 and 41 Communes in 2010). The trunk market defined in this way has in a second step been deemed to be competitive and fully deregulated. Not being part of the EU framework, the Swiss NRA did not need to notify the EC. If it would have had to, in light of the Portuguese case, the decision might have been vetoed for unstable market borders. It should be noted, however, that geographic segmentation of markets is the only legal tool for geographic segmentation of regulation available to the Swiss regulator as the Swiss framework foresees no flexibility of remedies. Imposing geographically differentiated remedies is therefore currently not a viable option for the regulator.

The **UK NRA** has been the pioneering NRA regarding geographic segmentation of markets in Europe. Its current WBA access regulation foresees three markets: 1) MDF areas where BT is the only operator present, 2) MDF areas where in addition two or more alternative operators with own infrastructure or over LLU are present[26] (or three when BTs market share is greater than 50%) and 3) areas where in addition four or more alternative operators are present (or three when BTs market share is lower than 50%). While market 3 is fully deregulated as it ensures competition, in market 1 full regulation including cost orientation and price-control (RPI-X) is imposed. Finally, in market 2 the additional price control remedy is withdrawn. Combining full deregulation with a segmentation of remedies between different markets, this proposal corresponds to date to the most flexible regulatory approach adopted in the EU. The European Commission reminded the NRA, however, that the sole criterion of the number of operators is not sufficient for geographic market segmentation, but that homogeneity has to be ensured checking for possible geographic variations in market shares and pricing. It invited the NRA in particular to provide additional structural and behavioural evidence, such as data on barriers to entry, marketing and sales strategies and service characteristics, which could further sustain the geographic market delineation. The European Commission did, however, not veto this decision and it was subsequently adopted.

Moreover, in the leased line market, the **UK NRA** in a detailed analysis in 2013 has defined geographic markets for high performance traditional interface terminating segments of leased lines (>8mpbs).

---

[25] IRG (09) 01 Phase II Case Spain 090116
[26] Presence means here a coverage of at least 65% of the MDF area.

Effective competition has mainly been found in the Western, Eastern and Central London area (WECLA). The WECLA has been slightly extended in the recent market analysis and follows 421 post code areas where competition is assumed, i.e. two or more competitors with own infrastructure and relatively low market shares of the incumbent. In practice the NRA estimates the number of potential competitors in a postal sector with a flexibility point within 200m of business sites. It is supposed that 200m can be reasonably bridged by any new installation of fibre to provide high performance leased lines services to a client. Then, the average number of potential operators per business site in the postcode sector was calculated and contingent postal codes with at least two alternative operators were grouped together. Market shares of the incumbent in this area were shown to be considerably lower and some geographic differences in prices have been detected. The only area with significant differences in economic conditions when compared to the rest of the country was given was then shown to be the WECLA. Finally, very high speed leased lines (622 Mbps) were defined separately (as a joint national market) as both submarkets seemed to be equally competitive. Regarding regulation, the NRA proposed to fully deregulate competitive markets (this is automatic) and to impose price control on the remaining markets. The European Commission has cleared this proposal and it was subsequently adopted.

It should further be notes that in some countries the low performance copper-based WBA market has been fully deregulated at national level (**Malta**, **Romania**). In Malta, in the retail market, two equally large competitors were found (incumbent and cable) and joint dominance could not be demonstrated in 2008. The assessment could possibly be different in light of NGA services and deployment today. In Romania, strong infrastructure competition seems to take place on a national level with the incumbent having relatively low market shares when compared to cable competitors. Also, competitive conditions were not judged sufficiently heterogeneous to warrant sub-national markets. The European Commission has accepted full national deregulation in 2010, but cautions the NRA to follow market development especially of competitive conditions across areas closely.

To sum up, even though the European Commission works towards a homogenous approach to regulation across Europe, current regulatory policy on geographic segmentation of regulation is highly fragmented. Various different approaches and criteria still co-exist. This may also be a result of the current absence of a veto of the European Commission on remedies.

As the review of relevant regulatory cases shows, in several cases a geographic segmentation of markets has not implicd full deregulation. On the other hand depending on the concrete details of regulation a remedy set can also nearly correspond to no regulation. The Austrian NRA in its WBA decision, for instance, had imposed only accounting separation in competitive areas[27]. Therefore, both the segmentation of markets as well as the segmentation of remedies may in practice imply near equivalent market regulation. The amount of fine-tuning then also depends on the preference of the regulator and the instruments it is ready to impose. The simplest form of fine-tuning would be cost-orientation and full deregulation. But different regulators have proposed different solutions than that including access-only obligations preventing foreclosure[28].

To conclude this overview, the approach of lighter regulation where competition is more intense is in line with theory when looking at static welfare – as increasing competition decreases the necessity of safeguarding competition. What is scarcely discussed are the detailed effects on investment incentives and the implementation of the remedies. For instance, LLU prices are usually based on uniform "cost-oriented" LRIC prices. As costs in many cases significantly differ across areas[29], uniform prices in regional markets may not set the correct investment incentives in all areas as will be seen in the next sections. Of

---

[27] This decision was only rejected by a national court.

[28] Generally, it can be noted that remedies in service-based markets could also be lighter as entry barriers are lower than for instance in the market for wholesale (physical) network infrastructure access.

[29] Ilic et al. (2009) show that in Switzerland costs can differ by a factor 6 across geographic cost clusters.

the reviewed cases there is one exception to this: The Dutch regulator has – under formally national regulation - imposed geographically segmented prices according to local capital expenditure requirements.

Geographic Segmentation of markets

Country	EC Case No.	SMP operator(s)	Product market	Geographic market	Geographic Segmentation of Remedies	Type of regulation imposed	Status	Comment on status
Austria	AT/2013 /1442-1443 AT/2009 /0932	Telekom Austria	Terminating segments of leased lines > 2 Mbps and < 155 Mbps	1) 12 competitive communes having population >15'000, more than three infrastructure based competitors and a market share of the incumbent <50% 2) The rest of the country	-	1) None 2) Access, non-discrimination, price control, accounting separation and transparency	Withdrawn (partially)	A first market definition of 12 cities and the rest of the country has been contested by the EC. The partial decision of regulating high speed lines in the rest of the country had been withdrawn. In its more recent fourth round market analysis (2013), the Austrian NRA reverted back to no geographic markets and uniform remedies. The EC vetoed this decision as there seems to be a lack of evidence for homogeneous competitive conditions across all regions in the country. BEREC has shared this view. A new final decision is still pending.
Czech Republic	CZ/2012 /1322	Telefonica CR (incumbent)	Wholesale broadband access	1) Districts with at least three infrastructures 2) Rest of the country	-	1) None 2) Access, non-discrimination, cost-orientation, accounting separation and transparency	Withdrawn	BEREC supported largely the NRAs proposal. However, the proposal was vetoed by the Commission and it has not entered into force.
Finland	FI/2013/ 1328-1329	27 regional incumbents	Wholesale broadband access	111 regional submarkets aggregating contingent municipalities with similar competitive conditions (number of competitors and market share of incumbent), 104 of which are non competitive and 7 competitive	-	7/111 markets: None 104/111 markets: Access, non-discrimination, and transparency	Adopted	The EC did not comment on geographic issues and the decision has been adopted.
Finland	FI/2013/ 1328-1329	27 regional incumbents	Wholesale (physical) network infrastructure access	111 regional submarkets aggregating contingent municipalities with similar competitive conditions (number of competitors and market share of incumbent) all of which are non competitive	-	111 markets: Access, non-discrimination, cost-orientation, accounting separation and transparency (small regional incumbents are subject to lighter regulation)	Adopted	The EC did not comment on geographic issues and the decision has been adopted.
Portugal	PT/2008 /0850-851	PT	Wholesale broadband access	1) MDF areas where Cable (at least 60% coverage) and one LLU operators are present 2) Other MDF areas	-	1) None 2) Access, non-discrimination, price control (retail minus), accounting separation and transparency	Adopted	The EC had raised concerns that in some competitive MDF areas the market share of the incumbent is still above 50%. It has therefore invited the NRA to carefully monitor the future evolution of competitive conditions, but the decision was adopted.
Portugal	PT/2010	PT	Leased lines	1) Competitive trunk	-	1) None	Adopted	The EC has states that the geographical

Country	Ref	Operator	Market		Remedies	Status	Commentary
	/1121		(trunk)	segments (between 110 local exchanges where two or more alternative operators are present) 2) Other, non competitive trunk segments	2) Access, non-discrimination, cost orientation, accounting separation and transparency		segmentation is primarily based on the number of operators and was insufficient. Given the important differences in market shares and network duplication the Commission did, however, not contest the decision.
Switzer-land	-	Swisscom	Leased lines ("trunk" segments, where trunk corresponds here to the a competitive backbone segment of the leased line market)	1) Lines between communes where more than 3 operators are present (25 Communes in 2009 and 41 Communes in 2010) 2) Other leased lines ("terminating" segments)	1) No regulation 2) Access, non-discrimination, cost-orientation and transparency	Adopted	This decision is in force. Not being part of the EU framework, the Swiss NRA did not need to notify the EC. If it would have had to, in light of the other cases, the decision would probably have been vetoed for unstable market borders.
UK	UK/2010 /1123 UK/2007 /0733	BT	Wholesale broadband access	1) MDF areas where the incumbent is the only operator present , 2) MDF areas where two or more alternative operators are present (or three when BTs market share is greater than 50%) and 3) Areas where in addition four or more alternative operators are present (or three when BTs market share is lower than 50%	1) Access, non-discrimination, cost orientation, accounting separation and transparency as well as an additional strict form of price control (RPI-X). 2) Access, non-discrimination, cost orientation, accounting separation and transparency 3) None	Adopted	The EC reminded the NRA that the sole criterion of the number of operators is not sufficient for geographic market segmentation, but that homogeneity has to be ensured checking for possible geographic variations in market shares and pricing. The EC invited the NRA in particular to provide additional structural and behavioural evidence. It did, however, not veto this decision.
UK	UK/2013 /1428 UK/2008 /0747-0749	BT	Terminating segments of leased lines 8> Mbps with tradition interface leased lines	1) WECLA: Areas with two or more alternative competitors with own infrastructure and low market shares of the incumbent 2) Rest of country	1) None 2) Access, non-discrimination, price control (RPI+X), accounting separation and transparency (for bandwidth at 622 Mbps no remedies are imposed)	Adopted	The EC has cleared this proposal and it was subsequently adopted.

Geographic Segmentation of remedies

| Austria | AT/2013 /1475 AT/2007 /0757 | Telekom Austria | Wholesale broadband access | 1) MDF areas with two or more alternative operators present, incumbent market share below 50% and serving more than 2'500 households 2) other areas | 1) Accounting separation 2) Access, non-discrimination, price control (retail minus), accounting separation and transparency | Adopted but rejected by national court | The EC had signalled to veto a first proposal of the NRA to introduce geographic markets in 2008. The NRA had then adapted its proposal to define a national wholesale broadband access market and proposed to withdraw most remedies in more competitive segments of the market. The EC had accepted this proposal. It was, however, rejected by the Austrian Administrative Court 2008 leading to an implementation of regulation |

Country	Code	Operator	Market	Geographic scope	Remedies	Status	Comments
Germany	DE/2010 /1116	Deutsche Telekom	Wholesale broadband access	National	Access, non-discrimination, accounting separation and transparency	Adopted	without geographical differentiation. In the recent fourth round of market analysis (2013) the NRA again proposes a national market, this time with nationally uniform light remedies (retail minus price control). But no geographical differentiation of remedies is proposed. A new final decision is still pending.
							While the NRA had analysed a possible submarket, national pricing strategies of the incumbent indicated a national market. The Commission agreed that there is no conclusive evidence for a geographically differentiated regulation. The decision was then adopted.
Italy	IT/2011/ 1230	Telecom Italia	Wholesale broadband access	National	1) Access, non-discrimination, cost-orientation (however a particular methodology leading to higher prices), accounting separation and transparency 2) access, non-discrimination, cost-orientation, accounting separation and transparency	Adopted (partially)	The market has been defined nationally. A concrete proposal on the geographical extent of the segmentation of remedies is still pending.
Netherlan ds	NL/2009 /0868	KPN/Reg gefiber	Wholesale (physical) network infrastructure access	National, but fibre LLU access prices are geographically segmented according to capital requirements (14 areas in 2013).	Access, non-discrimination, transparency and price control. For fibre: Max. Internal rate of return (IRR) allowed up to risk adjusted WACC + risk premium 3.5%. If exceeded price caps are reduced. Local fibre LLU access price caps currently range from 16-26€/month. There is also a cap for a national tariff of 18€.	Adopted	The EC has accepted the Dutch regulation proposal.
Poland	PL/2011 /1184	TPSA	Wholesale broadband access	National (earlier proposal : 11 cities, rest of the country)	1) Access and non-discrimination 2) Access, non-discrimination, cost-orientation, accounting separation and transparency (earlier proposal in 2011: no regulation in 11 cities, cost regulation in the rest of the country)	Withdrawn	Both decisions have been withdrawn, meaning that currently national cost-based regulation is still in place.
Spain	ES/2008 /0805	Telefonica	Wholesale broadband access	National	1) MDF areas where Cable and two LLU operators are present and incumbent has <50% market share 2) Other MDF areas	1) Access and non-discrimination 2) Access, non-discrimination, cost-orientation, accounting separation and transparency	Withdrawn	While BEREC had supported the NRAs view on both a national market and the possibility of geographic remedies in this case, the EC had asked the Spanish NRA to withdraw the proposal for insufficient evidence.

2.3.   Review of Literature

As described above the subject of geographic segmentation of regulation is receiving increasing attention of regulators as the mass market roll-out of new access infrastructures by the incumbent, but also new entrants at local scale, are increasingly requested by the public and taking place. A popular example described in regulatory practice section is the UK WBA market, where the regulator has first introduced geographic differentiation of regulation by essentially adopting full deregulation in areas where four or more alternative infrastructures are present and imposing differentiated regulatory remedies in areas where only the incumbent is present and in areas where two or more alternative infrastructures are present.

Some academic articles analyse the geographical impact of geographically uniform access prices (Lestage & Flacher (2010) or Flacher & Jennequin (2012)). To date, however, a comprehensive theoretical analysis of geographically segmented access regulation has been undertaken only by Bourreau, Cambini and Hoernig (2012b)[30]. In addition, De Matos & Ferreira (2011) analyse similar effects. In this section this literature will be summarized. An overview is presented in Table 3. The detailed effects of the different regulatory options according to the literature are described in the next section.

In Bourreau, Cambini and Hoernig (2012b), in a Greenfield setting two potential vertically integrated incumbent firms locally roll out own equivalent infrastructure with increasing fixed costs in more rural areas and with identical cost functions. Both operators can choose in which areas they will deploy own infrastructure and decide on the level of their investments, but they are supposed to start roll-out in densest areas first and roll out subsequently in ever less dense areas. While one operator can roll out in more areas than the other the possibility that operators deploy alone in different areas is not given.
In a static game in a first stage a regulator is setting the regulated wholesale access charges in all areas. In a second stage the two firms simultaneously and non-cooperatively set their investment levels. Then, a possible downstream entrant (and an incumbent in areas where only the other incumbent is present) decide whether to enter or not considering the access charge. The entrant chooses randomly an operator for access in case two incumbents are present. Finally, in a fourth stage all retail operators compete with horizontally differentiated broadband products for final broadband customers by setting possibly also geographically differentiated retail prices. The model uses quasi-linear preferences following Shubik and Levitan (1980) and an exponential investment cost function for the market model. Using this framework the effects of a variety of possible geographic regulation instruments are analysed. In particular the authors describe the effects of geographically differentiated access price regulation in areas with different cost levels and/or competitive conditions and geographically differentiated remedies.

Similarly, in the absence of legacy networks and assuming a fibre Greenfield market, in an endogenous entry setting, De Matos & Ferreira (2011) perform a market simulation with Cournot differentiated goods retail competition. It is assumed that two areas exist, one with low deployment cost and one with high deployment cost such to contemporarily exclude the possibility of infrastructure competition. In the first stage integrated and downstream operators decide in which markets to enter and in the second stage they compete on the retail market for end customers. The paper simulates the resulting geographic market structure and welfare.

While there are to date no other articles taking geographic regulation explicitly into account, some look at the converse problem: the impact of uniform regulation on geographic coverage considering geographic differences in cost levels. Lestage & Flacher (2010) in a similar static stage game as Bourreau, Cambini, et al. (2012b) assume Bertrand retail competition with vertically differentiated

---

[30] Pereira & Ferreira (2011) also consider geographic access prices. As the detailed functions of their algorithm is, however, not disclosed it is difficult to compare their model.

goods. In most of the paper the source of quality is assumed to be generated by the service provided on the infrastructure, i.e. duplication of access infrastructure is not socially valuable[31]. They then analyse the impact of uniform access price regulation on the geographic market structure and welfare.

In a setting with legacy technology and geographically uniform prices Flacher & Jennequin (2012) show that maximum coverage is reached without regulation but that this is not optimal. With one potential vertically integrated fibre incumbent and a potential downstream entrant as well as Cournot retail competition with vertically and horizontally differentiated goods it is shown that the social optimum is achieved in case the regulator sets not only access prices but also a coverage requirement[32].

Regarding the effects of regulation, the details of the imposed regulatory instruments matter. In European regulatory practice the debate on options to geographically fully deregulate or impose lighter sets of remedies is intense as the review of regulatory cases shows. On the other hand in academic research the analysis of welfare effects of geographically segmented regulated access prices or the problems implied by uniform pricing have not yet received much attention. In the next section the detailed findings of the existing papers with respect to the different regulatory options will be reviewed and put into perspective. The literature is summarized in Table 8 in the annex.

2.4. Review of regulatory options and effects

The different regulatory options to approach geographical access regulation identified by the literature include geographically uniform access regulation as well as competition and/or cost-based geographical segmentation of remedies and prices. Uniform access regulation is a regulation which does not foresee any geographic segmentation. Such a regulation may include any of the regulatory access remedies (access, non-discrimination, transparency, cost-orientation, price control) or none (full deregulation). In case price control is imposed, prices under uniform access regulation do not vary across areas. On the other hand geographical segmentation of regulation is a type of regulation where the detailed regulatory instruments may imply geographically different regulatory conditions. This includes the imposition of different regulatory access remedies in different areas, the imposition of access prices which vary geographically based on the level of competition and/or the required investment cost in a given area and full deregulation of some areas. The detailed effects of different geographic regulatory policy options in light of the literature under consideration are analysed in this section.

1) Uniform access regulation

Uniform access regulation describes settings where there is no geographic segmentation of regulation of any kind. Nevertheless, uniform regulation can have geographic effects on the market. Uniform regulation analysed in the reviewed literature include: full deregulation (free market), cost-based access prices and (any other) uniform access price regulation (e.g. maximising static and dynamic welfare). A particular form of uniform conditions in the access market is given by the case where no access products are available.

i) Uniform access pricing

Under **uniform above-cost access pricing** an access charge above marginal cost is set at the same level in all areas, independently of the level of competition or investment cost in these areas. This is a common case as current regulatory practice in Europe implies that remedies do not necessarily need

---

[31] The authors provide, however, an alternative specification where the source of quality is supposed to be driven by the underlying infrastructure. In this case firm B accessing a high quality infrastructure A is able to replicate its high quality services. With a possible own lower quality infrastructure this is not possible.
[32] Technically this would correspond to a beauty contest including minimum coverage requirements.

to be differentiated geographically even if competitive differences are present. It should be noted that long run incremental cost (LRIC) price regulation is also considered to be an above cost access price regulation as it applies a positive rent. Bourreau, Cambini and Hoernig (2012b) show that setting a high uniform access charge means that investment incentives increase both the extent of single infrastructure areas (SIAs) and of duplicate infrastructure areas (DIAs). The typical trade-off between maximising per area welfare of connected areas applying low access prices and increasing coverage to generate additional area welfare in marginal areas applying high access prices arises. It should be noted that this analysis assumes Greenfield investments and therefore the absence of a legacy network. This allows to abstract from migration effects which would in the context of this model likely lead to an excessive level of complexity.

Independently of how investment cost is specified the authors show with their market model that the social benefits from investing in duplication in a marginal area in case of uniform prices are negative. A regulator would therefore in this setting wish to decrease the investment incentives for duplication and therefore the extent of the DIAs with respect to the extent of SIAs, the reason being essentially the duplication of fixed costs in case of duopoly[33]. This can only be done by decreasing the prices in DIA areas relatively to SIA areas. Any uniform pricing (including cost-based pricing described below) is therefore never optimal and higher welfare can be achieved with geographically segmented regulation according to competition.

Lestage & Flacher (2010) show in a substantially similar game-theoretic setting as Bourreau, Cambini and Hoernig (2012b) that, when investment costs increase towards rural areas and two potential fibre incumbents - having an outside option with "traditional" low-quality technology - decide on investments, imposing (uniform) regulated access prices limits the area of total coverage and retail prices is reduced when compared to the free market. On the other hand, high differentiation of retail services can increase coverage. In addition, the authors show that access regulation limits the areas where both operators roll out as in Bourreau, Cambini and Hoernig (2012b). Subsequently it is shown that there are areas between DIAs and where no operator rolls out, where one operator rolls out in equilibrium (SIAs); but that for a subset of these areas there are two equilibria[34], where it is not clear which operator would invest and it is then uncertain whether there will be investment at all or not. This zone of uncertainty would only disappear in case the quality advantage between the firms is small. In addition, this zone is supposed to be moving towards more dense areas when the access price falls.

Also, Avenali et al (2010), while not directly modelling geographic effects, expect that geographically de-averaged access prices (above-cost in urban areas and at cost in rural areas) would raise welfare as this would induce more efficient investment in high density areas and low-density areas.

Cost-based access regulation is a particular case of uniform pricing as it would imply the uniform setting of prices at (nationally averaged) marginal costs. Fixed investment costs in the industry are typically very high (and varying across regions) and marginal costs are typically very low and do not differ substantially across regions. It can therefore be expected that a geographically differentiated cost-based access pricing (access prices per area set at marginal cost per area) would be equivalent to a uniform implementation. In Bourreau, Cambini and Hoernig (2012b), uniform cost-based access charges would reduce total coverage with respect to an unregulated setting. The trade-off between static and dynamic welfare does not arise here as SIA and DIA prices are set at the same level, implying that no additional profits can possibly be generated by an operator by investing in duplication. The duplicated infrastructure would have to be resold at marginal cost at wholesale level generating no additional potential wholesale profits while the potentially investing operator would already have access to the infrastructure at minimal possible cost from its rival to generate the same retail profits as without investment. Such regulation would be optimal only in cases where duplication is not feasible

---

[33] In case of high access charges and a high level of service differentiation further incentives for duplication would be necessary.
[34] Such areas are only present when the exogenous quality advantage of firm A over B is insufficient

(i.e. investment cost in a marginal area is very high). When duplication is feasible instead and cost-based regulation prevents it from taking place, this would correspond to a loss of welfare. Uniform cost-based access regulation is therefore not optimal. While duplication is not possible in this setting, it is instead with uniform prices above-costs as this starts to create wholesale profits for a duplicate infrastructure and lower opportunity costs for the second incumbent to invest. For this reason welfare would tend to be lower with cost-based uniform access prices than with uniform above-cost prices.

Lestage & Flacher (2010) similarly show that uniform cost-based pricing is not optimal as it is not taking into account the correct investment incentives and is reducing total and duplicate coverage even more than under uniform above cost access pricing. They also show that, when tastes are sufficiently heterogeneous, an optimal regulated access charge would depend on: the lowest investment cost across areas (or in other words the maximum population density), the lowest and highest quality valuations of consumers and quality (where the quality of the traditional copper network is assumed to be zero and the quality of new infrastructures strictly positive). Moreover, it is shown that the optimal access charge increases in the lowest investment cost across areas (and decreases in the highest population density).

ii)   Other forms of uniform regulation

There exist also other forms of uniform regulation, namely full deregulation and the case where wholesale access is not available (for example for technical reasons). In the case when access is not available the firms can make retail offers only where they have their own infrastructure. Firms then roll out both up to a point where per area duopoly profits become lower than the per area investment cost. Then one of the firms may roll out up to a point where per area monopoly profits become lower than the per area investment cost. While firms are symmetric ex-ante this leads still to differences ex-post, as in some regions only monopoly profits can ensure coverage and hence only one provider is present. Also, the case of full deregulation (when access is available) is a type of geographically uniform regulation. However, differently to the above cases it may imply geographically segmented commercial wholesale access prices. The market model of Bourreau, Cambini and Hoernig (2012b) shows that when services are sufficiently differentiated, downstream entry is beneficial to the industry due to a demand expansion effect (even though the retail profits of incumbents decline when giving access). In particular they show that in this case foreclosure (prices set to exclude the entrant) can never happen in SIA areas as giving access can increase overall industry profits with differentiated goods and the incumbent is able to extract such profit. In DIA areas, instead, foreclosure is possible only when there is low differentiation at the retail level. In such case the regulator could impose an access-only obligation preventing foreclosure, which might be welfare enhancing as will be also shown below. Finally, a regulator would only set regulated prices below the potential commercial wholesale prices.

Lestage & Flacher (2010) show with their model in a free, fully deregulated market, when considering two firms A and B, of which A always provides the higher quality service[35], that where firm A has rolled out infrastructure as a monopoly, it will not provide wholesale access to B, while a monopolist B would set a wholesale price such as to allow the provision of (higher quality) services by A. This, as excluding product A from the market would reduce the total profits possibly extracted from the market. Regulatory intervention is therefore necessary in monopoly areas to avoid foreclosure in SIA areas.
A different case is given when considering that the quality is driven by the infrastructure instead of services. Then, there is no more reason for foreclosure, as B can also provide high quality services when accessing infrastructure A. In this case, however, duplication is not desirable as infrastructure investment by B would only lead to a provision of the market with lower quality goods.

---

[35] Both quality A and B are considered to be preferred to a "traditional" outside quality which is provided in case of no investment.

From a practical point of view it might be interesting to consider company statements during the consultation of the BEREC Common Position on Geographic Aspects of Market analysis in 2008. Some local alternative infrastructure providers seemed to have a critical view on geographic deregulation leading to de-averaging of wholesale prices[36]. In the current situation the incumbent needs to charge a uniform wholesale access price (at a national price cap) and cannot offer lower prices as this would imply charging lower prices in rural areas as well. The price is in urban area therefore in practice also a price floor. These regulated prices could potentially be very high when compared to local urban DIA investment costs and deregulation in presence of any form of competition may potentially lower them decreasing the value of all infrastructures in the market especially of alternative investors[37].

2) Geographical segmentation of regulation

Geographical segmentation of regulation describes general settings where regulated conditions vary across areas. Regulatory instruments that can be segmented include regulated access prices according to competitive conditions and/or investment cost, as well as geographic segmentation of remedies in general - as for instance cost-based regulation in rural areas and "lighter" forms of regulation such as access-only obligations in urban areas.

i)   <u>Geographical segmentation of access prices according to competitive conditions and investment costs</u>

Bourreau, Cambini and Hoernig (2012b) describe pure geographical segmentation as optimal regulated access charges which are set in areas of different population density and therefore investment costs separately and which are differing in addition according to the competitive conditions in the area (SIA or DIA). No European NRA has to date chosen such a highly differentiated model with varying regulated access prices also according to the level of investment cost and there is currently no significant public debate. Also, No European NRA has to date implemented such a SIA/DIA distinction purely (rule of thumb) as it is assumed that the number of operators is not the only driver of competition. Furthermore, no NRA has proposed differing directly regulated access prices based on a single methodology according to competition. As has been shown the Dutch NRA offers, however, regionally segmented regulated NGA access prices.

While the authors indicate that such a type of regulation would offer maximum flexibility to the regulator and therefore lead to maximal welfare they also assume that it would be complicated to implement in practice as in-depth knowledge about local retail demand and cost structures as well as competitive retail market interaction would be necessary. Optimal regulated per area access prices would maximize per area welfare while ensuring that investment in the areas is viable (both for the SIA and the DIA case defining separate prices). As welfare in SIA areas decreases with the access charge, the SIA access charge is set just high enough to make an incumbent operator break even with its total area profits when investing. If its retail profits would be higher than the investment cost, the optimal access charge would be zero. The socially optimal extent of the SIA region is shown to correspond to the SIA region which would also develop when the operator could set monopoly access

<hr>

[36] http://berec.europa.eu/doc/publications/consult_erg_geo_markets_2008/fastweb.pdf

[37] Also, some regulators have imposed some form of **uniformity of retail prices.** Valletti et al. (2002) show that in the context of universal service a uniform retail pricing obligation is creating strategic links between areas that would otherwise remain unrelated. The paper shows that uniform retail pricing leads to lower equilibrium coverage of both incumbent and entrants. The effect depends also on the regulatory context of other universal service policies such as price caps or coverage constraints. For instance, in presence of a minimum coverage obligation the effect may be compensated, but the measure would lead to an increase of (uniform) prices. Anton et al. (1999), Choné et al. (2000, 2002) and Foros & Kind (2003) find similar effects. Hoernig (2006) arrives at similar results by stating that a uniform price imposed on the incumbent would reduce its coverage as it seeks to avoid duopoly entry. If imposed on entrants it reduces the incentive for duopoly entry and may lead to independent regional monopolies.

charges freely as it would extend its network as long as this is profitably possible too (i.e. the last covered SIA region would optimally have regulated access prices at monopoly level).

In DIAs on the other hand duplicative investment incentives exist as long as the investment cost in duplication is lower than the difference between expected DIA and SIA profits (for the incumbent, which does not provide access). Given the expected demand and cost functions and the SIA access charges, the socially optimal wholesale price can be calculated. If the incumbents DIA retail profits with respect to its SIA profits (when not being the access provider) are sufficiently high, the DIA access price can be set to zero, maximising static welfare while safeguarding investment incentives. A particular case is given when the SIA access charges are set at marginal cost. In this case investment in duplication would incur high opportunity costs in addition to investment costs which could not be compensated by any benefit. In this case, the optimal DIA charge would also be zero and investments in duplication would be unprofitable as no additional wholesale profits or additional retail profits could be generated. Duplication in this case brings no social benefits. In the market model used by the authors it is shown than duplication is optimal in no area when SIA access prices have been chosen optimally per area. Finally, it is shown that cost-based access prices (LRIC) per area are higher than the described optimal SIA prices as they include by design a positive rent (which is incompatible with a zero profit condition), and as it does not take into account retail profits. LRIC is therefore problematic even if it would be applied per area.

De Matos & Ferreira (2011) show in an endogenous entry market simulation with Cournot differentiated goods competition that geographically differentiated wholesale prices (areas are differentiated according to cost/competition) are socially optimal. At the same time, the authors state that in case of regional markets which are not independent implementation of geographic regulation becomes a highly complex task. Interdependencies may be justified by economies of scale and scope and network effects, or as will be shown later, by uniform (retail) pricing obligations. In particular, deregulation of the more competitive areas may trigger unexpected consequences such as a change to a monopoly situation in an adjacent market. They also show that therefore a deregulation of a subset of regions based on an "N-plus" rule of thumb (Xavier, 2010)[38] is not sufficient to guarantee that the introduction of geographic remedies is welfare enhancing.

The problem of interdependencies raised by De Matos & Ferreira (2011) is largely avoided by Bourreau, Cambini and Hoernig (2012b) by setting independent per area cost structures and by not considering network effects.

In Bourreau, Cambini and Hoernig (2012b) with duplication-based regulated access prices instead, different access charges in SIAs and DIAs are set, but the charges does not vary between areas with different investment cost requirements (or between providers). No European NRA has to date implemented such an approach purely (rule of thumb) as it is assumed that the number of operators is not the only driver of competition. Furthermore, no NRA has proposed differing directly regulated access prices based on a single methodology according to competition. Such an approach is less flexible than pure geographic remedies and therefore implies lower social surplus as optimally charges should vary also across cost clusters as has been shown above. Duplication-based regulated access prices have the advantage, however, to be more transparent and easier to implement for NRAs.

As before in Bourreau, Cambini and Hoernig (2012b) show that the effect of an increase of both SIA and DIA access charges on welfare is ambiguous. An increase in SIA access charges leads to a loss of static efficiency in the concerned areas, an increase in coverage and possible welfare gains from transforming SIA in DIA areas via opportunity costs. However, this last effect is positive only if increased competition outweighs the costs of additional investment. On the other hand, an increase in DIA charges would decrease static efficiency in DIA areas while also having an effect on the transformation of SIAs in DIAs via potential wholesale revenues. This last effect again is positive only

---

[38] Such are rule would foresee that the threshold number of firms below which regulatory remedies remain.

if increased competition outweighs the costs of additional investment. If this is not the case, then the regulator should set the DIA access charge to zero in order to limit duplication.

One feature of this analytical framework is that optimality conditions are such that there is a positive correlation between the socially optimal SIA and DIA access charge. Setting a very low SIA access price (increasing opportunity cost, lowering DIA investment incentives) would imply also lowering DIA access charges. This as low SIA access prices imply an already high per area welfare, meaning that the net benefit of extending the DIA area decreases substantially and that the regulator should reduce its incentives to invest in duplication by lowering also DIA access prices.

In equilibrium, finally, using the market model the authors find that optimal regulated SIA access charges are set above cost. DIA access charges, however, should be set above cost only in case of sufficient differentiation. Otherwise, the social benefit of duplication is insufficient to cover investment costs. Also, the market model predicts that optimal access charges in SIA regions are to be set higher than in DIA areas in order to provide investment incentives but keeping static welfare losses in DIA as low as possible. Also it is shown also in this case that (SIA) LRIC would not be optimal and tend to lead to too low access charges reducing welfare.

ii) Geographical segmentation of remedies

The European regulatory framework provides the possibility to impose a lighter set of access remedies in more competitive areas. From a legal point of view, this can be the consequence of a national market definition (with regional remedies) but theoretically also of geographic segmentation of markets. Popular examples of geographic differentiation of remedies may be the cited cases of the Spanish and Polish WBA markets where the NRAs proposed to lift cost-orientation in more competitive areas, imposing essentially only an access obligation to prevent foreclosure.

Bourreau, Cambini and Hoernig (2012b) assume that the regulator could maintain (welfare maximising) price regulation in SIA areas while imposing only an access obligation in DIA areas. Typically in this case wholesale access prices in DIA areas would then be freely negotiated. If the entrant feels, however, that the access price is exceeding a level that it allows to enter the market sustainably it may under the access obligation ask the regulator to impose a price based on a dispute resolution procedure. The regulator would then impose a DIA access charge lower than the foreclosing price and then set the corresponding optimal SIA access charge. With this procedure the incumbents would compete freely on the access price, provided it falls below the dispute settlement price.

Adjusting slightly the game setting (Bertrand competition with homogenous goods at wholesale level, where the entrant chooses the more convenient offer) Bourreau, Hombert, Pouyet and Schutz (2011) show that in an unregulated environment the softening effect makes the rival not providing wholesale access to a more aggressive retail competitor (setting lower prices) leading to multiple equilibria. The new wholesale profits have to be traded-off against possible losses of retail profits due to increased retail competition and demand expansion effects due to differentiation. This means that it is not always optimal to provide access, that undercutting at wholesale level is not always optimal and that the usual Bertrand result at wholesale level does generally not hold. It should be noted, however, that the softening effect disappears in case of full differentiation (i.e. independent goods) as then softening competition with relatively higher retail prices would not lead to higher wholesale revenues. When the softening effect is present though, Bourreau, Cambini and Hoernig (2012b) show that a low access charge implies higher profits for an access provider than for the rival which is not providing access. When the access charge is high enough, in turn, the contrary holds. This means that in a DIA setting there may be an access price below which giving access is more profitable than not giving access. Undercutting prices at wholesale level is therefore always an individual best response triggering a race to the bottom for providing wholesale services between the incumbents leading to marginal cost prices for both operators. In the market model used by the authors this equilibrium is unique when services are sufficiently differentiated and the expected dispute settlement prices are sufficiently low. If instead services are sufficiently homogeneous, the access prices of both operators will be set at the second equilibrium such that profits of providing or not providing access are equalised (and the access charge is above marginal costs). In this case no operator would again have an incentive to deviate. Finally, instead if the dispute settlement price is set sufficiently high, both incumbents may prefer not to make feasible offers (third equilibrium) but expect the regulator to set access prices hoping it will subsequently not be chosen for access provision. This is in particular the case when the expected dispute settlement price is higher than the access price that equalises anticipated duopoly profits with the profits generated ex-post when providing access in the DIA area at the profit maximising access prices (subject to the condition that the entrant is not foreclosed).

Finally, with both sufficient product homogeneity and a low enough dispute settlement price one firm offers a monopoly access price, while the other makes no feasible offer. An anticipated low dispute settlement price can therefore unexpectedly lead to monopoly prices.

Using the market model the authors then show how socially optimal prices could be enforced. If the socially optimal access charge is below the access price that equalises profits of providing and not providing access, the race to the bottom of DIA access prices must be stopped as strong competition has a too negative effect on investment incentives lowering welfare. The race to the bottom can only be stopped by setting a price floor at the socially optimal access price. If instead the socially optimal price is higher, it can in many cases be enforced by setting the dispute settlement price at the socially optimal price. In case, however, that the socially optimal access price is lower than the access price that equalises anticipated ex ante duopoly profits with the profits generated ex post when providing access in the DIA area at profit maximising access price (subject to the condition that the entrant is not foreclosed), this price cannot be achieved in equilibrium without further instruments.

Geographically segmented remedies can therefore lead to a socially optimal outcome. Whether this outcome is achieved or not depend on the details of how such regulation is implemented (especially for instance whether floors and caps are imposed). Overall this type of regulation seems to have similar informational requirements to the other approaches proposed to maximise local welfare.

2.5.   Conclusion

In the preceding sections the effects of geographically segmented regulation have been analysed in detail. Simplifying, the typical welfare effects of geographic regulation options that can be inferred by the existing literature are represented in Table 2.

	Static welfare (competition)	Dynamic efficiency (investment incentives in SIA and DIA)	Total welfare
Geographically segmented regulated welfare-maximising access prices according to investment cost and competition	Optimal	Optimal	**Optimal** (even if in the market model this implies no duplication)
Geographically segmented LRIC prices according to investment cost and competition (SIA)	Suboptimal	Suboptimal	**Suboptimal** (but better than uniform LRIC[39])
Geographically segmented remedies	Can be optimal	Can be optimal	**Can be optimal** (depends on mechanism)
Uniform/geographically segmented cost-oriented access price regulation (at marginal cost)	Suboptimal (but optimal in already covered areas)	Suboptimal	**Suboptimal**
Uniform above cost access price regulation (including LRIC)	Suboptimal	Suboptimal	**Suboptimal** (but better than marginal cost-oriented)
Uniform full deregulation	Suboptimal	Suboptimal	**Suboptimal**
Geographically segmented full deregulation	Suboptimal	Suboptimal	**Suboptimal**
Geographically segmented prices according to competition only	Suboptimal	Suboptimal	**Suboptimal**
Geographically segmented LRIC prices according to competition only	Suboptimal	Suboptimal	**Suboptimal**

Table 2 - Welfare effects of different geographic regulation tools

In light of the reviewed literature summarized above and the practical cases considered it is possible to draw conclusions for all identified regulatory options of geographically segmented regulation.

Geographic full deregulation

As has been shown, various NRAs have proceeded to full deregulation of some areas of the country (Austria, Finland, Portugal, Switzerland and UK). Different authors have argued that geographic (full) deregulation may lead to foreclosure. While Lestage & Flacher (2010) argue that this is possible even in SIA areas in case there is substantial quality advantage on the potential second incumbent, Bourreau, Cambini and Hoernig (2012b) argue that this possibility may be given, but only in DIA areas and only in case of low (horizontal) differentiation. Regulators should therefore use this tool with caution.

Geographical segmentation of access prices

In regulatory practice in Europe uniform above-cost access price regulation (e.g. LRIC) is still a commonly applied remedy (e.g. WBA in Sweden). The theoretical literature shows clearly that uniform access price regulation is no longer optimal, in particular in case of a roll-out of new infrastructures under geographically varying costs leading to geographically differentiated market structures. In particular local investment incentives are not sufficiently taken into account. Instead, Bourreau, Cambini and Hoernig (2012b) show that welfare optimizing prices would vary according to investment cost levels and competition and should be largely set be the regulator. No European NRA has to date,

---

[39] This is an interpretation and not demonstrated in the relevant articles.

however, used a coherent geographically differentiated access price model according to the level of investment cost and competition. A first step towards such an optimal solution is however made by the Dutch regulation. Geographically segmented fibre access prices according to investment cost (even though not according to competition) were defined resulting in access prices ranging from 16 to 26€ per month per unbundled fibre line depending on the cost cluster. Surprisingly the decision has to date received few attention regarding this particular aspect by other regulators in Europe, BEREC or the European Commission. It should be noted that such a regulation can be close to a solution which also differentiates prices according to competitive conditions as it is likely that in the urban areas where Reggefiber deploys such conditions may be rather homogeneous (cable competition). The question is then rather whether the price imposed by the authority is also welfare optimal.

Also regarding segmentation of regulation according to competitive conditions no pure SIA/DIA distinction has been adopted yet, as the European Commission judges such "rules of thumb" to insufficiently represent the level of competition. In light of the above regulators and researchers should consider increasing their efforts to evaluate possibilities to approach current access price regulation to a feasible form of socially optimal geographically segmented access price regulation. The benefits in case of success would be important. Today, for instance, higher uniform access charges would lead to both higher total coverage as well as more duplication. Regulators are therefore currently facing a trade-off on whether to increase such access charges to incentivise investment (e.g. with risk premia on top of cost-based regulated prices) or not. In such a situation regulatory action may well depend on the subjective preferences of regulators, or in other words on how much competition they are ready to sacrifice in order to induce investments in more rural areas. A regulator could for example also decide to only target static welfare (competition), by imposing marginal cost access prices. These preferences may be an additional driver of the state of broadband networks in European countries today, representing the result of past investments decisions (see Figure 1). When adopting an optimal regulatory regime of setting welfare-maximising SIA and DIA charges in all areas (such that SIA and DIA investment is viable and static welfare maximised), regulators would need to take into account the degree of product differentiation at retail level, investment costs and retail competition. Imposing optimal prices would lead to a total coverage which is maximal and to maximum static welfare per area (lower prices would mean that entry would not have been viable and welfare could not have been generated in the first place). When a geographically segmented access pricing approach could be adopted the regulators dilemma of trading-off static and dynamic efficiency would therefore be solved.

Reaching this objective seems a complex task and may require a long time for the development of appropriate regulatory instruments. It should be considered whether current regulation would not have simple options to make small steps in this direction.

In the framework of Bourreau, Cambini and Hoernig (2012b) it is shown that a local SIA LRIC price is not optimal as it includes a positive rent (and also the incumbents retail profits are not considered) and therefore it is higher than the price necessary to make local investment viable. Regulators, however, to date essentially use uniform LRIC prices. While Bourreau, Cambini and Hoernig (2012b) do not explicitly show this, their results can be interpreted such that local LRIC prices are leading to higher welfare than uniform LRIC. This is the case as in urban areas a local LRIC price would already exceed both marginal cost as well as a price that would make the investment viable (as it includes a positive rent. In case of uniform SIA LRIC prices the prices applied in urban areas will usually be much higher than local LRIC prices as a national cost base is considered. Therefore in urban areas uniform SIA LRIC prices would be such that welfare could be increased by decreasing the SIA access price towards local LRIC as the investment would continue to be viable and static efficiency would be enhanced. Conversely, in rural areas a uniform SIA LRIC price would likely exceed marginal costs but may in many cases be lower than the price that would make a SIA investment viable. In such areas an increase in the price could trigger investment and lead to higher welfare. In other rural areas, especially where investments have already taken place an increase of the charge towards local LRIC might, however, have the only consequence to reduce static efficiency. Overall, however, a scheme, which for instance would approach regulated price in urban areas to local LRIC while leaving the access charges in rural areas unchanged would be invariably welfare enhancing. Interesting this is largely corresponding to the practical implementation of the Dutch regulation, which foresees local tariffs in parallel to national tariffs. Regarding implementation the circumstance than in the Netherlands an operator can only choose one of the two tariff models may, however, distort this result and potentially lead nevertheless to welfare losses in rural areas. When compared to theory the regulator would then still need to develop a potential regulatory strategy for DIA cases. Overall, however, the Dutch approach seems to be largely supported by the literature.

Geographical segmentation of remedies

Recently introduced risk premia show that there is increasing awareness at the political level that investment incentives may be currently insufficient. However, a clear link of the extent of the premia to the dynamics of optimal local investment incentives is to date lacking and a significant debate on (partial) de-averaging of regulated wholesale access prices according to cost clusters seems still not to be taking place. Since 2008, however, several regulation proposals and decisions of member states not only of geographic full deregulation (as described above) but also of geographic segmentation of remedies have been observed leading in their result to (to some extent) geographically differentiated wholesale prices. The latter approach consists in practice mostly in imposing an access-only obligations in urban areas implying some form of retail-minus regulation, avoiding foreclosure of the entrant, and standard cost-based regulation in rural areas (e.g. Spain, Poland).

Given the informational requirements on setting welfare-optimizing geographically segmented access prices, Bourreau, Cambini, et al. (2012b) analysed whether a set of geographically segmented remedies can also achieve maximal welfare. In practice they proposed to largely deregulate DIA prices defining a dispute settlement procedure, which would prevent foreclosure of access seekers in case no viable access price results on the free market (corresponding to an access-only obligation). Foreseeing the market outcome the regulator would then need to set a welfare maximising SIA charges as well.

This type of deregulation may, however, have unwanted consequences. For instance (for sufficiently heterogeneous products) in DIA areas a race to the bottom for wholesale access prices may result in equilibrium. But too strong competition on the wholesale level may not be socially optimal as at some point investment incentives are reduced in a way to reduce overall welfare. Hence, there may be cases where a DIA access price of zero may not be socially optimal and the regulator should step in to prevent too strong wholesale competition setting a price floor at the socially optimal DIA price. As currently regulators still focus on maximising competition, this proposal is in contrast with current regulation. Furthermore, when the socially optimal access price instead is high (and above the DIA equilibrium price) it can be achieved in some cases by setting the dispute settlement price equal to the socially optimal price. In other cases also further instruments would be necessary. Overall it seems that there would be few cases when the socially optimal charge would be reached spontaneously on the market. While the regulator could add safeguards to ensure socially optimal prices (such as a price floor and cap) this would imply similar informational requirements as with geographic segmentation of regulated access prices. Regarding price floors it should be noted that to date no major practical case has received attention where access prices have been set by a regulated firm below the regulated dispute settlement prices. Even though this example seems not encouraging, regulators and researchers should try to further evaluate feasible dispute settlement processes able to lead to socially optimal prices.

To conclude, many issues still remain to be explored. Methods to approximate socially optimal SIA and DIA access charges and to implement them should be the focus of future research. Other subjects of interest may include the structural assumptions of the models explaining the effects of geographic regulation. For instance, only static settings are currently analysed and regulatory commitment could be a problem. Also, possible strategic links between areas due to scale and scope economies, network effects or uniform retail price obligations are not sufficiently considered. Moreover, a legacy infrastructure an investment sharing options should be present integrating the migration debate described in the introduction. Also, alternative competition models could be considered as well as endogenous entry in an extended theoretical model. Finally horizontal and vertical differentiation play a key role. The two alternative hypotheses of the source of vertical differentiation (service or infrastructure) in Lestage and Flacher (2010) indicate that researchers and regulators may still need to uncover the driving forces of innovation in the broadband market.

3. NGA Co-investment models

The roll-out of next generation access networks implies the largest investments in telecommunications since the beginning of the 20th century, when the copper telephone access networks were deployed by the state. In the preceding chapter operators were assumed to fully duplicate infrastructure when they would roll-out second NGA network. This is, however, not always necessary as operators can also invest jointly and share investment cost. This chapter will review joint roll-out possibilities and risk sharing agreements in general. In this introductory section, the extent of investment requirement is described and put into perspective.

Elixmann, Ilic, Neumann and Plückebaum (2008) show that single fibre[40] deployment costs are as high as 2'100€ per home connected (Table 3) even in a very urban cluster (Germany). There are, however, countries with substantially lower deployment costs in such areas such as Italy (1'160€). There are different reasons for this as differing construction costs across countries, differing existing duct and aerial cabling and their corresponding access conditions[41] and network topology. In particular, in addition, investment costs for in-house cabling are supposed to be higher in northern than in southern countries.

	Germany	France	Sweden	Portugal	Spain	Italy	Switzerland
FTTH investment cost (homes connected)	2'111€	2'025€	1'333€	1'548€	1'882€	1'160€	1'643€ (2'465 Fr.)[42]
FTTH investment cost (homes passed)	919€	930€	530€	776€	859€	504€	-

Table 3 – Fibre Greenfield deployment costs per home connected and passed, FTTH P2P (source: WIK)

Investment comparisons per home passed follow a similar pattern. Homes connected consider also costs that are incurred to activate a customer's connection which include in-house cabling, customer premises equipment and trunk cards[43]. The investment is in this case, distributed on an expected target market, i.e. 50% of the potential customer base[44] while for passed homes it is by definition distributed on 100%. Consequently, investment cost per home connected is higher than twice the investment cost per home passed. Overall, even in a small and dense country such as Switzerland full national coverage with a single fibre FTTH network would require investments as large as €14,3bn (connected homes)[45]. With 4.5m homes, this corresponds to a national average investment cost per home connected of around 3'200€ [46]. These high costs are again driven by the fact that connections become exponentially more expensive as population density decreases towards rural areas. Ilic, Neumann and Plückebaum (2009) show that in Switzerland in this case the last (very rural) cluster 16

---

[40] FTTH (P2P)

[41] e.g. in France it is assumed that operators may to some extent use existing infrastructure (sewer systems) reducing Capex costs significantly (increasing Opex though). The case in Italy is similar where ducts, covering about 8% of the population, used by Telecom Italia to deploy a CATV network between 1995 and 1997 (Socrate project) were opened to competition by the Italian Antitrust authority in 2001. Free duct capacity was in the past mainly used by Fastweb. In the case of Switzerland, the model assumes that incumbent overall digging costs are reduced by 20% by the possibility of using existing ducts[41]. In practice it should be noted that an EVU may save even a larger part of these costs as in many cases their duct networks have sufficient space left for a roll-out of an FTTH network.

[42] Ilic, Neumann and Plückebaum (2009) show that in Switzerland the investment per home connected in an urban area (the comparable cluster is cluster cluster 2) is 1'642€ per month. There it is however considered that FTTH would reach a market share of 75% and not 50% as this is more realistic in the Swiss case. Calculating a comparable value deployment costs in Switzerland would be around 2'000€ and therefore comparable to Germany or France. The exchange rate was assumed to be 1.50 Fr./€. When applying a more recent 2013 exchange rate (1.20 Fr./€) deployment costs would be comparably highest with around 2'500€.

[43] In the Swiss case in-house cabling is also included in homes passed.

[44] in Switzerland in the baseline model foresees 75%, the value in the table is adjusted to 50% though

[45] 21,4 Mrd. Fr.

[46] 4'800 Fr.

requires 10 times higher investments per access than the urban cluster 1 (around 1'320€). In the last cost cluster, then, it is shown that subsidies of around 11'000€[47] per home connected would be required to make an investment viable.

3.1.   Regulatory principles in Europe

Improvement of competitiveness of the duct market

In this section possible ways to reduce investment costs for any type of investor (single investor or co-investment partner) will be explored. In light of the monumental investment cost described a prominent question in the recent political debate in Europe was if there is anything that can be done to reduce investments required for an NGA and in particular a FTTH roll-out for all operators. The European Commission (2013) has recently published a legislative proposal to reduce the cost of rolling out high speed communication infrastructures in Europe. The initiative concentrates on civil engineering costs (i.e. digging up roads and lay down fibre) as around 80% of deployment costs seem to be associated with it. The European Commission hopes thereby to reduce investment requirements via efficiencies by 20-30%. The adoption by EU Parliament plenary vote is expected in the beginning of 2014.

The European Commission's proposals include the following specific measures:

   i)   *Telecoms operators should have the right to access the physical infrastructures of other network industries (e.g. electricity, water, sewage, transport) to deploy high-speed networks.*
   ii)   *Telecoms NRAs should be able to take binding decisions in case of a dispute and act as a single information point dealing with information on infrastructures and permit applications.*
   iii) *All newly-constructed buildings and those that undergo major renovation would be required to be equipped with "high-speed broadband-ready" in-building physical infrastructure.*

Essentially this proposal gives telecoms NRAs full control over the duct market. In practice, the draft regulation would firstly require all utility companies (such as electricity, gas, water, sewage, heating and transport) to meet reasonable requests by telecoms companies for access to their physical infrastructure in order to deploy high-speed networks. In the event that there are no legitimate reasons to reject the request (e.g. availability of space, security, interferences), the access seeking operator may request access at fair and non-discriminatory terms, i.e. at conditions and charges to be set (by default) by the telecoms NRA. Moreover, when performing civil works, companies which are partly or fully publicly financed would be required to meet reasonable requests from telecoms companies for coordination of and participation in civil works. Secondly, a set of rules is laid down regarding the access to information about these facilities. The minimum information which operators of such network must provide to a single point of contact operated by the NRA include *i) location, routes and geo-coordinates of the infrastructure, ii) the size, type and current use of the infrastructure and iii) the name of the owner of the infrastructure and a final contact point.* Applications for permits for civil engineering work for telecoms operators will be made over a coordinating single point of contact electronic platform operated by the NRA. Moreover, local authorities are requested to answer any request within six months. Thirdly, all newly-constructed buildings and buildings undergoing major renovation would be required to be equipped with high-speed broadband-ready in-building physical infrastructure. While it is unclear which technologies are included in this definition, it seems reasonable to think that traditional copper in-house wiring is excluded.

It can be expected that in many countries where such measures have not yet been applied, this proposal may lead to additional NGA investments using alternative duct infrastructures. As in many cases, entities operating duct infrastructures (other than telecoms operators) are publicly controlled – often by local authorities - and not necessarily operating in a profit maximizing environment, an access obligation can be reasonable in order to ensure potential entry in the broadband market via alternative

---

[47] 16'411 Fr.

network ducts. In addition, the proposal aims at increasing transparency and reducing bureaucratic costs. However, even if the potential investment cost reductions indicated by the European Commission are fully realised (20-30%) and single, duplicate and co-invested coverage is increased, the required investments in FTTH will remain very high and a profitable full coverage will remain unfeasible.

NGN co-investments

While the European Commissions' legislative proposal (2013) addresses generic possibilities to reduce operator deployment costs, cooperative investment may reduce investment cost further in case of a roll-out of more than one operator in an area. The most typical case would be in areas where two operators decide to roll-out fully in parallel[48]. With a joint roll-out and mutual access agreements the total investment incurred may be reduced substantially. Such a co-investment agreement, as will be shown, would not necessarily imply less flexibility for the operators or reduce competition.

NGA investment cooperations in Europe have been discussed by the NGA Recommendation of the European Commission which states that *co-investments and risk-sharing mechanisms should be promoted.* Such schemes are also analysed in BEREC (2012a)[49]. It is shown that to date there are few practical examples of co-investments in Europe and even less examples of interventions by regulatory or competition authorities on the conditions of such agreements. Cooperations have been registered only in France, the Netherlands, Portugal and Switzerland and they only account for a small portion of total FTTH deployments in Europe yet. BEREC (2012a) describes that NGA investment cooperations usually foresee two components. On one side the mutual access terms and on the other obligations regarding the roll-out, i.e. which part of the network an operator is responsible to construct and give access to to the other operator. In some cases such agreements are purely financial where one of the partners does not need to roll-out infrastructure or give access to its existing or future infrastructure at all. In case of joint-ventures, which is the strongest form of cooperation, investment costs and profits are shared under some rule and the entity would act independently, but as one single firm.

Both the European Commission (in an earlier draft version of the NGA recommendation[50]) and BEREC are concerned with possible limiting effects of such cooperations on competition. BEREC (2012a) notes that *"whether a market with more than two operators (e.g. three or four) may be compatible with competition depends however on numerous factors and in particular on the level of independence that these operators enjoy, especially within a co-investment agreement. While such a situation has to be assessed in detail in a market analysis or while national authorities may adapt more specific guidelines in this respect it may be said in general that if sufficient independence between the operators is ensured, a market with more than two, i.e. three or more, operators may under optimal circumstances raise low concerns about collusion and the competitive situation".*

Of the different sharing regimes considered the BEREC report assumes that sale of long term IRUs (indefeasible rights of use) on a fibre in a multifibre network to a competitor may be regarded as largely equivalent for it to controlling a fully independent own network[51]. Similarly to a case where infrastructure is fully duplicated, it is in the current regulatory framework possible that co-investments

---

[48] I.e. in separate duct systems

[49] A detail review can be found in annex 2

[50] The European Commission stated in annex III of the second draft of the NGA recommendation that to create sufficient upstream competition co-investment agreements need to be i) based on multifibre, ii) partners should have strictly cost-oriented access, iii) they must effectively compete downstream and iv) sufficient duct capacity must be installed. Also a sufficient number of access providers would be necessary (three or four). This draft is no longer available on the European Commission homepage.

[51] This view is shared by the EC in the NGA recommendation where it is stated that "multiple fibre lines allow alternative operators each to fully control their own connection up to the end-user. In addition access seekers can obtain full control over fibre lines, without risking discriminatory treatment in case of mandated single fibre unbundling."

lead to sufficient competition in the market for wholesale (physical) network infrastructure access to justify full deregulation of LLU (copper as well as fibre). Overall, it can be assumed that co-investment schemes may lower duplication costs and increase duopoly coverage, while having potentially negative effects compared to traditional duplication.

Multifibre deployment

Ilic, Neumann and Plückebaum (2009) estimate costs and potential network coverage under different scenarios. Compared to a single fibre network they explain multifibre networks and relevant cost drivers as follows:

- In-house wiring: The higher number of fibres implies the deployment of larger cables (depending on the number of fibres per home, e.g. four) and more splicing work at the building entry point.
- Drop cable deployment: In the drop segment of the access network (i.e. between the distribution and the building entry point) larger cables have to be deployed. Ducts, however, are here dimensioned in the model in a way that they could hold cables both in case of single and multiple fibre lines and there are no additional construction costs involved.
- Distribution point: Contrary to the single fibre case a distribution point where all operators have the possibility to connect drop multifibre lines has to be installed and every participating operator has to conduct splicing work.
- MPoP: In case of handover to the other operator at the more distant local metropolitan point of presence level (MPoP) and not at distribution point level, the network operating partner has to install additional feeder capacity and splice fibres through at the distribution point. This may imply constructing larger feeder ducts. At the MPoP the fibres also have to be connected to the respective optical distribution frames.

The additional costs for an operator to deploy a multifibre networks therefore depend on where the access point (splice closure) for alternative operators is installed. When compared to a single fibre network in the Swiss market Ilic, Neumann and Plückebaum (2009) estimate additional investment necessary for a multifibre network (before any interconnection of alternative operators) at around 12% (cluster 1) decreasing to around 2% (cluster 16) for handover at distribution point level (i.e. multifibre up to the distribution point). In case of handover at MPoP level (i.e. multifibre up to the MPoP) the additional investments required would be of 26% (cluster 1) and 12% (cluster 16). When considering the first six (urban) clusters, overall the multifibre model would imply around 9% higher investments in case of distribution point handover and 18% higher investments in case of MPoP handover. Intuitively, in rural areas the investment share of the drop segment increases (longer lines). As in the drop segment no additional investments for cables in case of multifibre are assumed to be necessary the relative additional investment for multifibre decreases towards rural areas.

What has to be considered also, however, is that once an operator is granted access to the multifibre network, it also has to invest in order to connect to the network. In case of distribution point handover, for instance, the alternative operator would need to duplicate investments to reach the distribution point. The estimated costs by Ilic, Neumann and Plückebaum (2009) are representing this, meaning that for a four fibre network and distribution point handover, total investment requirements increase with the number of cooperation partners connecting to the network. For instance in the first six clusters with distribution point handover the total investment requirement for a multifibre network increases by 21% (from 4'124 Fr. to 4'996 Fr.) with one cooperation partner (instead of none). Considering the above, the MPoP solution can be socially optimal in cases when multfibre backhaul is more efficient than duplicate network backhaul. In fact in the Swiss case even though there was an extended debate on this, several cooperation partners agreed on handover at MPoP level. The cost estimates of Ilic, Neumann and Plückebaum (2009) are broadly in line with other estimates of Polynomics (2009) which estimated additional costs of 10% for multifibre networks and of the Swiss incumbent Swisscom estimating additional costs of 10 to 30%, depending on the case considered. Considering the above a possible national multifibre obligation as discussed in Switzerland might therefore raise costs also in monopoly areas reducing typically total coverage to some extent. These additional costs to society need to be traded off against benefits.

The European Commission's acknowledges the potential of multifibre in its NGA Recommendation (2010) stating that multifibre has several advantages and may be conducive to long term sustainable competition. It is stated that multifibre

- *allows partners full control of their own connection up to the end user*
- *enables an end-user to subscribe simultaneously to several service providers connected at the physical layer, which could in turn help develop new applications;*
- *facilitates churn, since no manual cross-connection operation is needed at the concentration point, any churn request may be dealt with without any down time*
- *lowers operating costs when compared to a single fibre FTTH scenario;*
- *ensures that access seekers can obtain full control over fibre lines, without risking discriminatory treatment in case of mandated single fibre unbundling.*

The main use for the customer in urban areas is therefore that a multifibre dose is installed at the home which allows potentially to choose one or more physical access provider simultaneously and easily switch between them (in Switzerland for instance four fibre connectors are installed). Cases where more than two operators are chosen simultaneously seem to date, however, rare in the Swiss market.

Coverage

Ilic, Neumann and Plückebaum (2009) assume a fixed average revenue per user of 57€ per month independently of the service purchased (single, double, triple play)[52] and independently of the number of entrants. For Switzerland, it is then estimated that traditional fibre infrastructure competition, i.e. investment in two independent parallel networks, would be profitable in this case for up to 16% of households. Using multifibre co-investments it is estimated that this coverage can be increased to up to 54% of households[53]. Surprisingly, even four operators would be economically viable under these assumptions for 36% of households[54]. These results are, however, assuming certainty of (symmetric) market shares after investment. As such certainty is not given in reality the actual coverages may be significantly lower. Finally, (maximum) total coverage under these demand assumptions is given by the potential profitable coverage by a single operator roll-out (single fibre) at around 60% of households (corresponding to 8.3% of the national territory)[55]. In the model of Ilic, Neumann and Plückebaum (2009), it is therefore predicted that - even in presence of cable - about 60% of the population could profitably be covered by an FTTH network (single fibre) and that for a very large part of these accesses (54%) physical FTTH infrastructure competition on the basis of a multifibre co-investment is viable[56].

3.2.   Regulatory practices

While co-investments can lead to operators having a comparable level of independence as in the case of a fully parallel roll-out, this is not necessarily the case. BEREC (2012a) distinguishes two forms of investment cooperations. On one side long-term cooperation agreements are considered where no common company is founded and access agreements are made for instance on a single fibre

---

[52] Assuming 35 CHF for single play (telephony), 65 CHF for double play (telephony and broadband), triple play 80 CHF (telephony, broadband and IPTV) and business 252 CHF, and applying services shares of 15%, 16%, 51% and 9%, an average ARPU per connection of 85 CHF results.
[53] 43% when handover is done at distribution point level instead of MPoP
[54] 16% when handover is done at distribution point level instead of MPoP
[55] In case of single operator multifibre roll-out 54% of households (in both the MPoP and the distribution point scenario).
[56] In the WIK model multifibre cooperations and costs structures do not affect total coverage.

infrastructure or also under indefeasible rights of use (IRU) on dedicated fibres in case of multifibre. On the other side the authors consider joint ventures, where the companies take equity stakes carrying jointly the full financial risk of the investment and reselling wholesale products jointly to the shareholders as well as possible downstream outsiders.

Long term cooperation agreements

Typically, many co-investment cases observed to date in Europe have foreseen limitations to independence and flexibility of participating operators. The following horizontal agreements part of NGA multifiber long-term cooperation agreements had for example been notified under objection proceedings to the Swiss competition commission[57]:

- Layer 1 exclusivity (notified in all major cities) foreseeing a clause whereby a partner commits not to give access at layer 1 to third parties

- Compensation mechanisms (notified in all major cities except St. Gallen), foresee that from a certain degree of usage of the network a transfer payment between the partners is necessary

- Investment protection clause (or non-discrimination of the partner) (notified in all major cities), foresees that access products cannot be offered at lower prices to third parties than to the partner

- Information exchange clauses (notified in all major cities except St. Gallen)

The Swiss competition authority has found that all these clauses (with the exception of information exchange) could potentially restrict competition. Such a finding could still be confuted by sufficient competition in the market (wholesale physical network infrastructure access and wholesale broadband access). However, in both markets, restricted to only fibre and including both dedicated and shared fibres, significant market power was found, especially for the technical problems making it difficult for cable operators to directly enter the market for wholesale physical network infrastructure access. Indirect effects through the retail market were supposed not to be sufficiently strong given that the only operator able to offer LLU on national level is supposed to be Swisscom. The competition authority was therefore unable to exclude an intervention in case the operators would agree and implement these clauses. Most clauses have subsequently been cancelled by the operators. BEREC (2012a) show that it is essential whether the investment cost is shared upfront or whether there are subsequent usage based charges transforming via the legal instrument of the co-investment potentially fixed costs in marginal costs manipulating competition. This is possible both in the case of long term access agreements as well as under joint venture.

Unlike in Switzerland in France cooperation agreements are largely defined ex-ante by regulation. Consequently there is less space for intervention of the competition authority. Essentially, the French regulation foresees that any firm wanting to roll-out FTTH in an area consults the market (via NRA) for interested firms in layer 1 co-investments[58]. If there is interest by other operators to participate in such an investment, multifibre is rolled out (at least one fibre per co-investor) and the operators essentially participate bearing equal shares of the investment cost for the multifibre infrastructure between the home and the distribution point[59]. In exchange, they receive a long term indefeasible rights of use (IRU) which define access agreements largely equivalent to property.

Independently of whether the roll-out took place by a co-investment or not, infrastructure operators in France must then provide (ex-post) access at reasonable and non-discriminatory terms to unbundling products at the distribution point. Differently to the co-investment such prices include a risk premium. This applies to very high density areas (i.e. communes with more than 250'000 population, where at

---

[57] See BoR (12) 41, Wettbewerbskommission (RPW2012-2)
[58] ARCEP Decision 2009-1106 of 22 December 2009
[59] ARCEP Decision 2009-1106 of 22 December 2009, article 3

least 20% of the houses consist of more than 12 units[60]). There, the distribution points are set for houses with more than 12 units directly inside the building. Similar terms apply in non dense areas[61] where, however, the distribution point is much more distant (such as to collect more than 1'000 lines). The NRA therefore imposes a larger extent of shared network outside dense areas. As an example France Telecom and Free have signed an agreement in July 2011 where 5 million households should be reached outside very-high density areas by 2020. Legally the French approach is interesting as it regulates fibre access in a symmetric way (i.e. applied to any firm on the market independently of the competitive situation). Also, in Portugal Optimus and Vodafone both construct own independent NGA networks in different cities. An agreement foresees mutual access.

Joint Ventures

Structural joint ventures of multiple telecoms operators in Europe are rare. In this case operators jointly control a company and divide investment costs and profits. In Holland a Reggeborgh-KPN joint-venture rolls-out an FTTH network. KPN, Reggefibre and other operators then buy access to layer 1 products from the joint-venture at regulated prices. The price caps are differentiated according to cost (capex) levels ranging from 15.52€ to 25.99€ per month in 2013 (14 different areas proposed). As described in the earlier chapter these prices are the result of a DCF model taking into account cost and demand over the lifetime of the investment (the regulated price sets the net present value to zero).

Also, under a proposed joint venture in Fribourg in Switzerland (Swisscom-Groupe E) other horizontal agreements have been rejected[62] by the competition commission. In this case, the agreement would have foreseen that ducts would remain under the control of the respective partners and that non-discriminatory wholesale offers are made. The competition authority had, however, ruled that the agreement would not constitute an independent new unit on the market taking over relevant assets of the partners – so-called full function joint venture - and considered therefore only the horizontal agreements. The main agreements were:

- The joint ventures layer 1 access price[63] is fixed over the whole term of the contract (same for co-investors as also third parties) in the agreement
- There is a minimum order quantity for layer 1 products (same for co-investors as also third parties). I.e. small alternative operator could not provide sufficient scale and would not be served by the joint venture.
- The operators fix a common price for access to their ducts (which remain under their respective control).
- Both operators make bids to the joint-venture indicating total roll-out costs per area. A clause foresees that the costs taken into account – bid of the winning operator – are increased by a fixed agreed mark up.
- The operators commit to not compete with the joint-venture operations at later stage
- The sale of layer 1 access products at the building entry point to third parties is restricted

The authority has shown that all these clauses could potentially reduce effective competition in the market for wholesale (physical) network infrastructure access. In June 2012 the joint-venture has adapted the clauses according to the decision of the authority. In order to ensure full coverage of the region, the Canton was requested to enter the capital of the joint-venture. The Cantonal Government had agreed to do so. At the same time Swisscom has decided to abandon the project and the cooperation form is now similar to the other Swiss agreements. Finally, in Italy Trentino NGN (controlled by the district authority) and Telecom Italia have set up a joint-venture where it would roll-out in dense areas (70%), while Trentino NGN would roll-out alone in the rest of the area.

---

[60] The decision states some further conditions for definition
[61] ARCEP Decision 2010-1312 of 14 December 2010
[62] Swiss Competition Commission. Case 41-0623: FTTH Freiburg
[63] terminal segment, i.e. from the distribution point to the home.

Structurally as will be seen under a joint-venture the partners can control the access costs of all downstream players. Under (long term) access this is not the case, as the incumbent may always retain access at marginal cost.

3.3. Review of Literature

The essential question the literature explores is the effect that different regulated and unregulated co-investment options have on investment, competition and welfare. As is the case with the applied regulatory work on the subject, theoretical and empirical literature essentially distinguish joint-ventures and (long term) access agreements. The key feature of a joint-venture is that the roll-out may be undertaken jointly and that the partners maximise joint profits and set a single downstream access charge for the partners (and a possibly different one for outside operators). While such agreements are generically considered to be co-investment agreements, it is not entirely clear which types of access agreements should be considered co-investments. In an access agreement, the (local) network remains under full control of an incumbent which gives access at a price possibly above marginal cost. In this an asymmetry in the market is created as the investor active on the downstream market may face only its marginal network cost upstream. It may consequently in these cases be impossible for the operators to reach efficient monopoly allocations as under joint-venture. In theory any above marginal cost access price may create additional rent (an investment contribution) for the investor supporting its investment. While many types of access options are considered by the co-investment literature, only the subset of these agreements including an ex-ante fixed investment contribution are usually considered to be co-investments[64], as in this case the investment risk can be equally shared. This section will, nevertheless, compare all joint-venture and (long-term) access options analysed.

Most of the co-investment literature considers (ex-ante contracted) joint-ventures. One particular form of joint-ventures is when insiders can access the infrastructure at marginal cost (access price set by the regulator or by the partners), where the network therefore can be used freely after the investment has taken place. Typically such a configuration would lead to intense downstream competition. Cambini and Silvestri (2013) call this basic investment sharing[65]. Also, in addition to these broad categories of cooperation an intermediate case is considered. The access innovation literature considers the case where the joint-venture maximises joint profits setting a jointly optimal investment level, but where the competitor would not enjoy marginal cost access as the incumbent paying above-marginal cost (regulated) prices.

Regarding access agreements instead, a broad range of options is considered. Essentially, access charges can be fixed (independent of quantity) or linear or nonlinear in quantity (e.g. fixed plus a usage base charge together or a usage based charge with quantity discounts). Ex-ante is considered, as usual, to consist of contracts signed before the investment takes place, while ex-post contracts are signed afterwards. Fixed charges can be optional (i.e. are effectively paid only when access is actually requested, which may not be the case when demand turns out to be low ex-post) or non-optional (to be paid in any case). In addition charges can be unconditional or conditional on the market outcome and in particular the level of demand in case of uncertainty. All these access options can refer to prices on the free market as well as to regulated prices (e.g. LRIC, FDC or marginal cost). In addition

---

[64] This seems in line with the definition given in the NGA recommendation: *Co-investment in FTTH means an arrangement between independent providers of electronic communications services with a view to deploying FTTH networks in a joint manner, in particular in less densely populated areas. Co-investment covers different legal arrangements, but typically co-investors will build network infrastructure and share physical access to that infrastructure.*

[65] Usually in one way or the other marginal cost is born by the partners. Be it via the joint-venture or via own NGN marginal costs equal for both operators.

to the mentioned co-investment and access options often a benchmark case is considered where no access is possible.

Essentially, the literature shows that co-investments can extend duopoly (and sometimes total) coverage but risks reducing competition. As welfare effects are therefore contradictory, the social desirability of a particular co-investment depends on the fine details of the co-investment agreement and the outside option to which it is compared: for instance, whether both operators have non-discriminatory access to the infrastructure built, the regulatory environment, downstream competition, uncertainty, risk aversion, the structure of the access charges and the amount of investment required. Unsurprisingly, theoretical conclusions depend on the hypotheses assumed. It will be shown, however, that nevertheless conclusions and recommendations to date are largely consistent. The following section will provide an overview of the literature based on one basic paper (taking also into account geographic aspects) described initially. Table 9 in the annex summarizes the NGN co-investment options considered in the literature and the main assumptions and results of the respective papers.

3.3.1. Co-investment under NGN regulation

The most detailed analysis of co-investment (basic sharing) to date is provided by Bourreau, Cambini and Hoernig (2013). The authors use a similar model than Bourreau, Cambini and Hoernig (2012b) considering next to geographic effects also uncertainty[66] and access to outsiders once the investment has been undertaken. However, unlike most other co-investment papers, the authors consider a Greenfield investment and therefore no migration effects from copper to NGA reducing model complexity (and practical relevance) to some extent but defining a good starting point for further analysis. While most other articles consider access regulation an alternative scenario to a co-investment scheme, this article considers the two simultaneously.

Regional incumbents can here decide on the extent of Greenfield NGN investments in their respective home areas. They invest up to the (most costly) area where gross profits can just cover the investment cost. They then announce their plans and can decide to what extent they would like to co-invest in the home area of the other incumbent - where investment cost would be split and access granted at marginal cost. This as the authors assume that higher internal access costs reducing competition would not be tolerated by the regulator (largely corresponding to the regulated ex-ante co-investments proposed by the French regulator). The paper also assumes that the co-investors then set jointly a local access charge to the co-invested infrastructure for the outsiders seeking access. The paper analyses the investment incentives for both total and duplicate/co-invested coverage that a co-investment option creates in three market regulatory environments: no access (benchmark), traditional regulated (NGN) access and the free market (in duopoly areas only).

When only (regulated or commercial) co-investment options exist (i.e. the outside option is that no access is granted), the only way to provide NGN products is by having access to their own infrastructure (be it via single roll-out, duplication or co-investment). In the case when the competitor can somehow share investment costs and then access the technology at marginal cost, as under duplication, operators would earn duopoly profits in the areas concerned (which are reduced compared to the profits in monopoly areas). The only difference being that the investment cost can now be shared, reducing the cost for duplication and extending the duopoly coverage (which is usually lower than the monopoly coverage) when compared to a case with no access option. Duplication would therefore be fully substituted by co-investment and the duopoly coverage extended. In line with the rest of the literature, which will be described, the paper concludes that usually total coverage is not affected by co-investment options. This might be case only when co-investment duopoly profits exceed monopoly profits, i.e. when a joint roll-out would lead to efficiencies reducing the total investment cost or when there is a strong demand expansion effect. Reasonably the former is not the

---

[66] but assuming risk neutrality

case. For instance Schneir and Xiong (2012) show that additional investments would in reality be necessary in case of any co-investment, as infrastructure would need to be more flexible and necessitate more equipment to be able to host two partners (even when considering a relatively economic passive optical network (PON) FTTH infrastructure[67]). Regarding the latter, as in most other papers, differentiation is key. If goods are sufficiently differentiated the sum of gross profits of two active firms may despite increased competition be larger than the profit of a monopolist. When this effect is sufficiently strong to balance the likely increase in investment cost, an increase in total coverage might theoretically be possible when introducing the co-investment option (even when introducing only the possibility of duplication). In addition, it is shown that when the probability of low demand increases, not only both monopoly and duopoly coverages are reduced but also the difference between the two, meaning that a co-investment scheme would also reduce coverage risk.

When instead traditional regulated ex-post[68] access (uniform linear usage based fee both in monopoly and duopoly areas) is also granted and demand is high, partners would ask for access outside of co-invested areas, i.e. in all areas where only single infrastructure is deployed. It is assumed that access is not asked for in case demand is low and that then profits would be the same as under no access. Here, it is assumed that also downstream entrant can enter on the retail market based on access regulation (both in single as well as co-invested areas) but also only in case of high demand. In such a case it is shown that usually an increase in the access charge increases both single and co-invested coverage. Then with respect to the no access case regulated access usually undermines investment incentives (total coverage) unless the regulated access charge is high and product differentiation too. Secondly, the introduction of regulated access is now an alternative to the co-investment creating an opportunity cost for a co-investors which reduce co-investment coverage (in the extreme case of access at marginal cost, there wouldn't be any incentive to co-invest anymore independently of the investment cost). When deciding on whether to provide regulated access (instead of no access) to co-investors the regulator therefore has to trade-off enhancing competition in single infrastructure areas with a reduction of incentives for co-investments, reducing infrastructure competition. The authors argue that a solution could be that regulated access is not provided to co-investment partners (only to downstream entrants), but this may not be feasible from a legal and practical point of view.

Finally, investment incentives are analysed under voluntary access, where in co-invested areas due to infrastructure competition access prices are fully deregulated (regional regulation) while traditional regulation remains in place in single infrastructure areas. In this case the co-investors will allow local access only when profitable, thereby weakly increasing their local profits. Co-investment coverage therefore increases with respect to both the no access as well as the regulated access scenario (while voluntary access has usually no effect on total coverage as regulation in monopoly areas remains in place). This effect is essentially due to the fact that investors here have full flexibility to maximise their profits.

Voluntary access for co-investments is, however, not necessarily socially optimal, as it may lead to higher retail prices. The authors show that such deregulation of co-investments only provides higher welfare than no access in case services are sufficiently differentiated. Also, compared to regulated access, voluntary access only leads to higher welfare when services are highly differentiated and the compared access charge under regulation is high. The first result is obtained as the introduction of a freely and jointly profit maximising access charge by the co-investors may be used to soften downstream competition[69]. This may increase the co-investors total profits even in presence of a new entrant when compared to no access implying, however, less welfare. In the case where instead goods are highly differentiated, there would be no such negative competitive effects of deregulation

---

[67] PON allows to passively bundle the traffic of multiple fibre lines on one single backhaul line, reducing feeder costs, but potentially limiting flexibility.
[68] i.e. access is asked for after the investment is sunk and demand uncertainty has resolved
[69] This is also described in BEREC(2012), where it is stated that compensatory mechanisms after the investment which imply effective above marginal cost access prices can be strategically used to reduce in the market.

and welfare would be enhanced. The welfare effects of voluntary access compared to regulated access are then straightforward. Given sufficient differentiation (negative effect of co-investment on competition is weak) and a high enough access charge under regulation, local welfare in a deregulated co-investment area is higher than in a regulated single infrastructure area. Also, as has been shown, voluntary access would increase co-investment coverage. Therefore, only when differentiation is strong and the compared regulated access charges high may deregulation of co-investments be a socially better choice than traditional access regulation. The French authority seems to share this view as it has actually not only regulated co-investment access conditions but also ex-post access conditions to the infrastructure. Under the current regulatory framework, it may propose to lift this part of regulation when the co-investment grants sufficient competition therefore limiting negative effects on welfare.

3.3.2. Co-investment models as an alternative to NGN regulation

While the rest of the literature does not take geographical aspects explicitly into account, different aspects of the preceding model are also analysed when considering the presence of a legacy network on a whole considered, possibly urban area and upgrade investments which, depending on their size and the ability of the operators to sell quality services, may unlock additional willingness to pay.

Overall modelling approaches in the rest of the literature vary strongly. For instance, Nietsche Wiethaus 2011, Cambini and Silvestri (2012) and Cambini and Silvestri (2013) compare different exogenous risk sharing agreement options (traditional joint-ventures and basic sharing) to - alternative - traditional NGN regulation options (LRIC, FDC, marginal cost, free market, no access). Unlike Bourreau, Cambini, Hoernig (2013) these authors consider an incumbent with an existing copper network to which all players have non-discriminatory access at marginal cost (regulated). Except for Cambini and Silvestri (2013), these papers take into account uncertainty. We will now review the rest of the literature, considering the following broad category of models: Presence of uncertainty, differing ability of partners to sell NGN based products and the presence of outsiders. Subsequently the access innovation literature is analysed where access conditions between the incumbent and the co-investor may differ and the NGN investment has no quality effect exclusively reducing access cost. Then, the literature on long term access regimes is reviewed under which the incumbent continues to fully control the network, while still being able in some cases to share risk. Finally, the empirical literature on co-investments is described.

a) Certainty

The simplest overall setting is provided by Cambini and Silvestri (2013) which consider a given roll-out area under certainty. Consumers' willingness to pay for NGN depends on the amount of investments and the two considered possible incumbents are equally good in transforming quality investments in willingness to pay. They then rank market outcomes regarding investment, competition and welfare for the traditional joint venture case, the basic sharing case as well as the traditional regulated monopoly case. Cambini and Silvestri (2012) introduce also uncertainty making similar but more detailed conclusions considering in addition the case where NGN is left unregulated, while the legacy network is continued to be regulated. Nietsche and Wiethaus (2011) consider a similar model under uncertainty comparing the basic sharing case to specific regulation such as LRIC or FDC.

In Cambini and Silvestri (2013), a downstream competitor has the possibility to enter a basic sharing agreement with the incumbent before the investment (ex-post access in case of agreement is granted for free for the partners, having to pay only their marginal costs for NGN). Duplication is therefore excluded. Investment costs as well as possible wholesale profits are as usual equally divided. Consumers are here having demand for basic broadband which can also be offered based on the legacy network and one for value added services based on NGN as in Foros (2004) and Katz and

Shapiro (1985). How much the NGN investment increases the consumers' willingness to pay depends on the industry's ability to transform input quality improvement into output.

Essentially, two scenarios are analysed. One where all operators are part of the co-investment agreement and one where there are outsiders asking for usage based access ex-post. In the regulated scenario, Cambini and Silvestri (2013) assume that no type of investment sharing option exists and that the regulator sets the welfare maximising access price to the incumbents infrastructure (ex-post and linear usage based) for all access seekers. It is shown that in this case the optimal NGN access price is set at marginal cost (as for copper). The investment extent would then depend on the willingness to pay for NGN services and investment costs and it would decrease with the number of outsiders using access, as these would compete and reduce industry profits (Cournot). In equilibrium in the basic sharing scenario, instead, when all firms participate in it, industry profits and investment incentives are increased compared to the regulatory scenario as now also NGN profits generated by the co-investing (former downstream) competitor can be taken into account when making the investment decision. In this case the whole spill-over of the investment on the competitor can be considered when deciding on investment. Typically any other form of collaboration (e.g. ex-post access, especially when regulated) would reduce the amount of rent that can be extracted from the competitor reducing investment incentives as will also be shown in Inderst and Peitz (2013). Finally, in case of a traditional joint-venture, when firms are also free to choose the access price to the co-invested network, competition can also be softened increasing profits and investment incentives even further.

Equilibrium output it is shown to be highest under basic sharing. Firstly, it is higher than under joint venture, where partners may set a high access price to dampen downstream competition restricting output[70]. Secondly it is higher than under regulated access, even though access prices to the network are identical in equilibrium (marginal cost), as investment and therefore demand are increased under basic sharing. Finally, output under joint venture in equilibrium would usually be higher than under regulation (at least when willingness to pay for quality investments is sufficiently high and costs sufficiently low)[71].

It is also shown that the ranking with respect to total welfare in this model is identical. Increasing both investment and competition, basic sharing is superior to access regulation (similar conclusions will be described in Nietsche Wiethaus (2011) and Cambini and Silvestri (2013)). By contrast, a joint venture with freely chosen access charges is a combination between strongest investment incentives and strongest restriction of competition. Again, when willingness to pay for quality investments is sufficiently high and costs sufficiently low it is shown to be superior to regulation as in this case investment is having more welfare value. Finally, a joint venture option is shown to always generate less welfare than basic sharing as the increase in investment incentives in this model with an exponential cost function and Cournot competition can never compensate the loss in terms of competition.

b) Uncertainty

Nietsche and Wiethaus (2011) and Cambini and Silvestri (2012) introduce uncertainty such that willingness to pay is enhanced only in case of success. Conversely in the case of failure willingness to pay is not enhanced. The binary nature of success allows to introduce the element of uncertainty without unduly extending the complexity of the model. In Cambini and Silvestri (2012) differently to Cambini and Silvestri (2013) and Nietsche and Wiethaus (2011) and following more closely Foros

---

[70] . As stakeholders will be redistributed 50% of the JVs profits the access price would only have a financial impact on an operator when its use of the infrastructure would be different than 50%/50%; this is not the case in this symmetric and certain environment. However, otherwise a JV is vehicle for internal transfers (similar to full compensation payments in the Swiss case under loose cooperation agreements).

[71] The ranking in terms of consumer welfare is identical.

(2004) the willingness to pay for quality of consumers may vary across firms. The results found under uncertainty are not in contrast with the results found in Cambini and Silvestri (2013) under certainty.

Differing ability to increase willingness to pay of consumers across firms

In Cambini and Silvestri (2012) again an incumbent with access to a legacy infrastructure has an option to invest in NGN under different possible exogenous regulatory regimes or a sharing option with a competitor. Demand is revealed only in the retail competition phase. Similarly to Cambini and Silvestri (2013) three access regimes are considered: Basic investment sharing, NGN regulation, and free NGN market. In all cases there a regulated copper option continues to be available.

The incumbent and possibly the alternative operator in case of co-investment must decide on when to invest in a given (supposedly urban) area under consideration (investment extent is supposed to be 100%). Investment costs are assumed to decrease over time[72] meaning that the investment is becoming more profitable over time and that at some point investment would take place. Practically, an exponential discount factor (between 0 and 1) is applied to a (quadratic) investment cost as in Bourreau and Dogan (2005) and Riordan (1992), depending on the adoption date of the new technology. The earlier the investment takes places the higher the discount factor, and consequently, the investment costs that need to be incurred to upgrade the network[73]. The investor will decide therefore on the investment timing, which will determine the investment costs. Until the moment of adoption the incumbent makes profits based on its legacy copper network. The NGN generates profits afterwards. The regulator in this model sets access prices ex-ante, but access prices can be conditional (i.e. higher in case demand turns out to be high). In this model, it is mostly assumed that the entrant has to commit to an access regime and cannot switch back to copper after demand is revealed. It therefore bears risk as well.

Under traditional NGN access regulation, it is shown that when the incumbent is much more efficient in creating willingness to pay for NGN services compared to the competitor, the regulator would set an expected welfare maximising price excluding the competitor from the NGN. This case is, however, assumed to be unrealistic. When the ability of the competitor increases slightly but the incumbent is still better than the competitor, the regulator would set an above marginal cost fibre access charge making its entry viable. Finally, when the ability of the competitor further increases and is only slightly lower than the ability of the incumbent, and when it is even higher, the regulator would set a negative access charge in case of success in order to incentivize the alternative operator to offer NGN based products, given that only its presence may unlock (quality) competition and possibly increased willingness to pay for NGN downstream. Negative access charges are, however, excluded and it is assumed that in such cases the fibre access charge would be set at marginal cost as copper. The authors also show that a situation where the regulator cannot set conditional access prices would be suboptimal, as the alternative operator could be inefficiently forced out of the NGN market in case of failure. This as above marginal cost NGN prices would be valid also in the case of failure and could imply that profits would be lower than with copper. Finally, the incumbent decides on the investment timing. The authors find that the better the competitor is on the fibre market the later the incumbent would invest (as in Foros (2004)). This occurs when the NGN access price is set at marginal cost, meaning that the investment is pure spill-over but also in the case of above marginal cost NGN access prices. Also, when the probability of success increases, the investment is undertaken earlier and the incumbents' incentives to invest decrease less strongly with the ability of the competitor.

In the case of full deregulation of NGN a simple take-it-or-leave it offer is considered as opposed to Nash bargaining considered below by Nitsche and Wiethaus (2011). Moreover, it is assumed that in

---

[72] This means that in this model there is always investment at some point.
[73] When investment takes place in period 0 the investment cost is not reduced at all. When taking place in period three it would be reduced by around 99%.

case of failure, the access price would be set at marginal cost. It turns out that the incumbent would set the NGN access price in case of success such that

i) when the competitor is significantly less efficient in offering value-added services, it is excluded from the market
ii) when the competitors' ability increases but not up to a point where he would be significantly better than the incumbent, the incumbent will charge above marginal cost prices which just allow the alternative player to enter the NGN market.
iii) when the competitor is considerably better, access is granted fully extracting the willingness to pay the incumbent would be unable to generate himself (monopoly prices).

These conclusions differ from Foros (2004), where the outside option is market exit (instead of copper) and the incumbent would charge an unconstrained NGN access prices excluding the entrant, whenever the competitor has a lower ability to sell NGN services. Here, the cases of exclusions are reduced as to make entry of the competitor viable even if it is (to some extent) less efficient than the incumbent. This is due to the trade-off that if it would not allow the entrant on the NGN it would continue to compete for basic services over the legacy network at regulated marginal cost access prices, which is creating an opportunity cost for the incumbent. Granting NGN access, the incumbent can at least earn some upstream profits, which if would not earn in case the competitor would continue to use the copper network. Lifting copper regulation would therefore substantially weaken the competitor's position. Finally, in case of deregulation of NGN the authors find that the better the competitor is on the fibre market the earlier the incumbent would invest as here the incumbent can always capture part of the rent of the competitor.

Under basic investment sharing, the two firms choose the investment time to maximize their joint expected profits. In equilibrium, when at the start the competitor is better than the incumbent (or when the incumbent is better but not too much), the investment is undertaken earlier when the competitor becomes better in selling NGN. Conversely when the incumbent is considerably better than the competitor an increase in the competitors ability would delay the moment of investment. This scenario is therefore representing an intermediate solution with respect to deregulation of NGN and regulation as it internalises the effects of retail competition.

The authors conclude as the rest of the literature that basic sharing leads to more (or the same level of) competition and output than in case of NGN regulation (but also than NGN deregulation).The equilibrium in terms of time of investment depends on NGN access conditions and therefore on the firms respective abilities to sell NGN in the retail market. The investment is undertaken earliest in case of deregulation, while the ranking between NGN regulation and basic sharing depends on the parameters. When the regulated NGN access price is set to zero (marginal cost), the investment is undertaken later than under basic sharing as in this case investment costs can be shared. When the regulated NGN access price is positive instead, the relationship is ambiguous. Intuitively, while in case of investment the competitor may always profit from some spill-over effect, the incumbent may in case of deregulation also capture a part of this rent via the upstream market. In case of NGN regulation instead – if the incumbent has not a considerably higher ability to increase willingness to pay for NGN - the regulator would set prices at marginal cost decreasing the incumbents' wholesale profits to zero. Investment incentives are therefore reduced and investments take place later. Finally, when the success probability increases the investments are in all scenarios anticipated. Uncertainty is therefore a major source for suboptimal investment.

The interpretation in terms of total welfare of this model is unclear. When the competitor is better than the incumbent in providing NGN services (and a regulator would consequently set the NGN access price to zero), basic sharing is always the socially optimal choice. Even though investment incentives are lower than under NGN deregulation basic sharing more than compensates this in with the intensity of competition. Also, NGN deregulation is more efficient than regulation in this case. When instead the incumbent is better (but with the competitors' ability not so low to be excluded) NGN regulation continues to yield lowest welfare, while the ranking of basic sharing and NGN deregulation is unclear. When the ability of the competitor is further reduced, the incumbent excludes it from the NGN market in case of NGN deregulation. In this case basic sharing is better than NGN regulation from a welfare point of view, while the relationship between NGN deregulation and basic sharing is ambiguous.

Equal ability to increase willingness to pay of consumer across firms

Nietsche and Wiethaus (2011) use a similar but simpler model as Cambini and Silvestri (2012). The factor which transforms quality investments in willingness to pay in case of success is assumed to be one for both the incumbent and the competitor. When access to NGN is granted, both players are therefore supposed to be equally good at selling NGN products. Again the outside option is regulated copper access. The regulatory options considered are now detailed regulatory regimes. Under LRIC, the access price is considered to be an average investment cost per unit (marginal costs such as the cost of production and distribution are again sustained in addition by both the incumbent and the competitor). It is assumed, however, that if the investment is unsuccessful and no additional willingness to pay is created by the NGN investment, the NGN LRIC access price is set to zero. Therefore, only in case of success can the incumbent pass-on investment costs to the competitor under LRIC. In case of failure, the willingness to pay of consumers is not increased and the incumbent would continue to sell copper products under conditions as before and could not recoup its investment cost. Under fully distributed costs (FDC), instead, access prices are also defined as investment cost per unit. But here the incumbent is allowed to recoup costs also in case of failure (i.e. positive regulated NGN access charge also in case of failure). The form in which investment costs are recouped in case of failure can be by a forced full switch to fibre or by continued parallel services, whereby, however, copper based products have to contribute to cover the NGN investment cost. Finally, a basic sharing agreement is considered as well as a deregulated NGN environment. In the case of NGN deregulation negotiation for access to the network in case of success is modelled differently to Cambini and Silvestri (2012) as a Nash bargaining solution is assumed, meaning that rent extraction by the incumbent is more limited.

Overall the authors show that in case of success competition is strongest in case of basic sharing where implicit access prices are lowest (in particular when compared to LRIC). As the equilibrium outputs in case of failure would be the same, overall expected quantities are increased with basic sharing. Moreover it is shown that LRIC leads to higher expected output than FDC, as the outcomes in case of success are equivalent, but as FDC would increase access costs for the competitor also in case of failure leading to lower output in this case. Finally, it is also shown that basic sharing generates more output than NGN deregulation as the latter leads to positive transfers in case of success.

When looking at investment, given that output in this setting is always symmetric, with LRIC in case of success investment costs are effectively reduced by 50%. With FDC the entrant bears this share also in case of failure. Under basic risk sharing instead all investment costs are entirely sunk and do not allow any allocation of investments as second stage marginal access costs leading to a high level of retail competition and consequently limited investment incentives[74]. Basic sharing therefore induces less investment than both FDC and full NGN deregulation. The ranking between basic sharing and LRIC, however, is not entirely clear. As under LRIC the incumbent has to share the benefits of the network in case of success, but it cannot recoup or share investment costs in case of failure, the investment incentives strongly depend on the probability of success. In case of certainty for instance LRIC would provide better investment incentives than risk sharing. In case the probability of success is low enough (under 85%), however, basic sharing, turns out to induce more investments as it allows to share not only benefits but also investment costs upfront.

---

[74] The authors, however, admit that risk sharing may also have other forms allowing for such transaction and improving this trade-off.

The authors finally compare the performance of these regulatory options in terms of consumer welfare. It turns out that risk sharing is superior to LRIC both in terms of competition than in terms of investment incentives. This is, however, not always the case with respect to other regulatory options such as deregulation and FDC. In a numerical example, the authors show that usually expected consumer surplus for a large range of parameters (probability of success lower than 90%) is highest for risk sharing, followed by FDC, deregulation, and LRIC. The high performance of risk sharing is due to its property of leading to a very high intensity of competition, but at the same time giving reasonable investment incentives allowing sharing of both benefits and costs in all cases. It should be noted that risk sharing remains optimal even if the probability of success is above 90% and in a certain environment. In this case only the ranking between NGN deregulation and LRIC becomes unclear. Interestingly FDC dominates both NGN deregulation as well as LRIC. Apparently the higher investment incentives more than compensate lower competitive intensity. Furthermore, with some uncertainty even NGN deregulation appears to dominate LRIC (for a large set of parameters). This final result depends on the particular form of access prices under deregulation (Nash bargaining) and the hypothesis of competition.

c) Outsiders

The only paper next to Bourreau, Cambini and Hoernig (2013) to consider ex-post outsiders in case of co-investment is Cambini and Silvestri (2013). In this case the insiders are able to set a (usage based) access price for outsiders which is potentially different from the insider fee. Results are however, not directly comparable as in Cambini and Silvestri (2013), as for the presence of a regulated legacy network option which changes the model fundamentally. Similarly, though the presence of an outsider undermines investment incentives, in particular in case of regulation.

When an outsider is considered, in the basic sharing case the partners continue to access the infrastructure at marginal cost while the outsider has to pay a higher NGN access fee. The outsider also has the alternative possibility to use the copper network at regulated marginal cost prices (same as NGN) or to not enter at all. Given the demand structure, the more the partners invest in quality the less attractive is providing copper services for the outsiders. Depending on the extent of investment, the outsider may therefore be fully excluded from the market even though access to copper is regulated. In equilibrium the authors show that when willingness to pay for quality investments is sufficiently high and costs sufficiently low, the partners set an external access fee so high, that the entrant is excluded from the NGN network. Intuitively when the competitive advantage from fibre over copper services is large the temptation to exclude the entrant from the NGN is higher for the partners as profits in such a situation increase. It is also shown that under the same circumstances the partners choose an investment extent in the preceding stage which is high enough to exclude the entrant also from entering via copper (even though access is regulated at marginal cost), in which case the investment level is identical to the one under no access. When willingness to pay for quality investments is instead sufficiently low and costs sufficiently high, the partners set an above marginal cost access price which makes entry viable. One of the reasons the entrant is not excluded in this case is that it is simply not fully excludable when regulated copper access is granted at marginal cost and the willingness to pay cannot be significantly enhanced at reasonable cost. Once the entrant is not excludable, access can also be granted to the NGN, where more rent can be extracted.

In the joint-venture case the partners instead choose the internal as well as the external access fee freely. When the willingness to pay for quality investments is sufficiently high and costs sufficiently low, the partners again exclude the outsider from the NGN via its access charge. In this case they would set their internal access charge at marginal cost in order to be able to compete at best on NGN base with the copper-based competitor. As before, however, in equilibrium the entrant is excluded also from copper based via investment extent when it is excluded from NGN based entry. In the converse case, the partners would set an outsider fee above marginal cost which would make NGN based entry for the competitor viable as well as an identical internal fee to overall soften (NGN) competition. The regulators intervention may here in both cases prevent discrimination and possibly foreclosure. It would again choose marginal cost access for all operators (insiders and outsiders), in which case the equilibrium investment under joint-venture would be the same as under basic sharing. French regulation is largely in line with this observation as it foresees ex-post access for outsiders but includes a risk premium.

Under a joint-venture the partners are again able to increase profits by reducing downstream competition. With outsiders, however, also under basic sharing some dampening of competition via the outsiders' access fee is possible. This means that for a given investment extent output is highest and investment level lowest under regulation (uniform regulation at marginal cost). Also output under basic sharing is higher than under joint-venture. The rankings compared to the no outsider case is now different as the presence of an outsider implies that the insider fee is set low by the partners. In equilibrium Cambini and Silvestri (2013) show that with an outsider, sharing agreements increase investments incentives (even more under joint-venture than under basic sharing) over regulation but dampen competition further and lead more likely to exclusion. However, the benefits are such that total welfare is always enhanced by sharing models over the regulated case. The exact ranking between basic sharing and joint venture is unclear and depends again on the willingness to pay for NGN and investment costs. It seems therefore that notwithstanding the fact that sharing agreements can lead to a reduction of competition and potential foreclosure of outsiders this can be socially optimal when compared to a situation with NGN regulation at marginal cost which would reduce industry profits with every outside entrant. Regulators fears of a reduction of competition are therefore well founded when outsiders are present. Nevertheless they should consider that such regulation can reduce investment incentives to a point where welfare is decreased.

3.3.3.   Access innovation

Some interesting insights can be obtained from the literature on cooperative access innovation. Mizuno (2009) considers access innovation representing investments with the effect of exclusively reducing network access costs[75]. While two firms compete à la Cournot with horizontally differentiated goods at the retail stage, in the investment stage, two exogenous options are considered. On one hand a non-cooperative regime in which the first moving incumbent alone determines the investment level maximizing its profits, and on the other hand a cooperative access innovation regime (joint venture) whereby the investment is chosen that maximizes joint profits of the incumbent and an entrant while sharing the fixed cost somehow and continuing to compete downstream. Unlike all other articles considered, the access fee the competitor has to pay ex-post is different from the one the incumbent bears. It has to continue to pay a usage based (linear) access fee which is set by the regulator.

Under uncertainty in a benchmark scenario an unconditional regulated access price is considered which does not adjust to realized costs and is fixed. In this case, the investment incentives for access innovation are higher in case of no cooperation, as the entrant does not have any spill-over from the access innovation (results are reported in Table 4). Even worse for the entrant, the access innovation will lead to increased competitiveness of the incumbent reducing its market share and profits. In a more realistic scenario where the regulator imposes a conditional cost-based access pricing rule, the access charge is a fixed multiple (usually above 1) of the realized access cost ex-post (e.g. adding common non-traffic dependent cost elements as a fixed percentage of the access cost on top). Expenditure for access innovation investment may be also included in this perspective. Under any such access rule now access innovation and cost reductions by the incumbent also have a positive spill-over effect on the entrant as a reduction of the access cost also reduces the access charge and therefore the entrants marginal costs.

---

[75] For simplicity it is assumed that the incumbents' marginal costs are equal to average costs.

When the spill-over effect (and access charge) is very small, the entrants' access costs are reduced much less than the incumbents, leading to a strong competitive imbalance, given that the entrants' costs increase relatively to the incumbents'. In this case the entrant overall does not benefit from access innovation and access charge reduction and it would in case of cooperation work to reduce investments in innovation. A non-cooperative investment by the incumbent would therefore lead to higher investments. When the spill-over effect becomes high enough, the entrant also benefits sufficiently from access innovation and cooperative investment can increase investment over non-cooperative investment. Finally, when the spill-over effect becomes very high, the entrant benefits more from the access innovation than the incumbent, whose overall benefits from access innovation may become negative due to competitive effects. No investment would then be undertaken as the entrant is supposed to be unable to invest alone. Access charges that are too high, therefore, contrary to intuition, do here not incentivize investment in access innovation but deteriorate it. This is however, only the case because of the particular regulated access price structure (fixed access rule). When the access charge is not a multiple of the access cost, but instead is set as a two part tariff, where non-traffic related costs are set separately as a fixed "set up" fee in addition to usage based charges, the scheme would represent a mix between a fix committed price and a marginal access cost rule implying that the limitation of investment incentives under the non-cooperative scheme are limited. The author suggests that regulators should therefore take care when structuring regulatory access products as incentives for both non-cooperative and cooperative access innovation can be distorted. Regarding the cooperation scheme per se it is not always effective but in case of a regulated (cost) conditional access charge it allows overall to enhance investment incentives. Also, given the above an increase in competition (goods becoming closer substitutes) reduces the range of regulated access charges (and spillover effects) for which cooperation is viable.

	Cooperative access innovation	Non-Cooperative access innovation
Fixed usage based access charge	Lower investment incentives	Higher investment incentives
Linear ex-post contracts with high spillovers	Higher investment incentives	Lower investment incentives
Linear ex-post contracts with low spillovers	Lower investment incentives	Higher investment incentives
Standard LRIC	Higher investment incentives (Higher total welfare)	Lower investment incentives (Lower total welfare)

Table 4 – Investment incentives from non-cooperative and cooperative access innovation

In the rest of the paper, the authors conclude that the usage based regulated access charge, considered the only instrument of the regulator, should be set below marginal cost in order to compensate for presumed market power at retail level both in the non-cooperative and cooperative regime. When the access pricing rule is such that the access charge is equal to realized incremental access costs (e.g. LRIC) the level of spill-overs are shown to be "large". It is then shown that such an access pricing regime would not only imply that cooperation leads to more investment incentives with respect to a non-cooperative regime, but also that under cooperation total welfare would be higher. In case of a two part tariff, it is shown that it might lead to higher investment incentives under the non cooperative scheme but also that it would not be welfare optimal in this context.

3.3.4.   Long term access agreements

a) Certainty

The co-investment options considered in the rest of this survey foresee joint profit maximisation. There are however, also possibilities to share investment risk without joint control. This is in particular the case with long term access agreements where a competitor may reach an agreement with the incumbent which foresees, for instance, a fixed unconditional ex-ante investment contribution in exchange for favourable ex-post access. Inderst and Peitz (2012a) as well as Inderst, Kühling,

Neumann and Peitz (2012) analyse the effects of different access options including ex-ante long term access agreements to NGA in a certain environment. They derive critical levels of investments below which investment is undertaken under different access options. The outside option is again represented by regulated copper access at marginal cost.

Two operators are supposed to fully control a hinterland of particularly loyal customers beyond reach for the competitor and served exclusively. In addition, non-captive consumers are located on a Hotelling line with uniformly distributed customers and products are located at the two endpoints. In such a setting it is shown that the equilibrium price difference of the two products increases with differences in consumers' gross utilities (or willingness to pay), marginal costs or the extent of hinterlands. It is, however, assumed that customers' gross utilities only differ between the firms when they use different technologies. With given fixed, price independent hinterlands - and therefore industry demand - the authors note also that the property that firms can only set a uniform price for all customers (captive and non-captive), means that firms with a larger hinterland are less aggressive in the competitive segment holding a lower market share in this segment. In this analysis, however, symmetric hinterlands are assumed. Analogously equilibrium conditions are derived for the case when demand in the monopoly hinterland segments is price dependent (as well as consequently industry demand). The NGN investment decision takes place consisting in a 0-1 decision in a regional market (the incumbent deciding first on investment). In the following the different network access scenarios for the competitor are analysed under certainty (for a summary see Table 5).

	Investment incentives (total coverage)	Investment incentives (duplication)
No access possibility	Lowest	Maximum
Ex-post: Linear access charge	Intermediate	None
Ex-post: Nonlinear access charge (full bargaining power with investor)	Maximum	Maximum
Ex-post: Nonlinear access charge (not full bargaining power with investor)	Intermediate (equal or higher than with linear access charge)	Intermediate
Ex-ante contract option (co-investment)	Higher than corresponding ex-post option	None

Table 5 - Effect of access options under certainty (case of price independent demand)

When no access possibility for the competitors exists, duplication may occur if investment requirements are very low. In the other extreme case, investment requirements are so high that not even a single operators' investment is viable. In the intermediate case, only one of both firms' investments is viable and only one firm invests in equilibrium. As a second option traditional ex-post access is considered. It is first assumed that access fees take the form of a linear charge per subscriber to recoup the investment and that the investor has full bargaining power.

When industry demand is price independent, an increase in linear access prices above marginal cost is shown for the competitor to work like an increase in its marginal costs and leads to an equivalent increase in the retail price in equilibrium (see De Bijl and Peitz (2006)) as the whole marginal cost increase can be passed on one-to-one in equilibrium. The entrants profit remains therefore unchanged with changes in the level of the access charge. It is further shown that in equilibrium the same is true for the incumbents' prices via opportunity costs.

The incumbent would therefore be the only firm benefiting from this access price increase being able to extract rent from the entrant via higher wholesale profits. Foreclosure never happens in this case as the investor is able to always increase its profits through access, extracting rent generated by the entrant (competitors' hinterland). Total coverage is therefore increased with an access possibility. Investment incentives are, however, not efficient here as the linear access charge determines jointly the level of industry profits and their distribution between the access seeker and the investor. Under this scheme the competitors' net profits from access are the profits generated in duopoly at retail level (above wholesale cost). Duplication and a possible reduction of the competitors' access cost to marginal network costs[76] would not impact the retail profits of the two firms, most importantly leaving the competitors' total profits unchanged. Duplication at any positive investment cost is therefore never possible in such an environment. In addition, a change in the distribution of bargaining power has here no effects as the competitor is indifferent about the level of the access charge.

When ex-post non-linear access prices are considered, for instance, not only a usage based charge has to be paid by the access seeker, but also a fixed charge. Compared to the linear access charge, more rent extraction would then be possible. As in a joint-venture, the usage-based access charge would then be chosen high enough to set marginal cost conditions such to maximise industry profits (monopoly outcome), while the fixed fee would allow the participants to divide the profits according to bargaining power between the two firms (in case of full bargaining power, extracting the entire additional profit, being largely equivalent to a joint venture). A two part tariff option therefore increases the investment cost that can be borne by the investor and investment incentives for total coverage when compared to standard linear access charges. When a shift in bargaining power is considered, it has no direct effect on the market outcome, but on the distribution of rents (and indirectly on level of investment). When not all of the bargaining power is with the investor, rent extraction and total coverage are lower. Regarding duplication, when the entrant does not invest on its own and uses access it has zero profits under nonlinear access in case of full rent extraction. The decision on when to invest in duplication is for the entrant then equivalent to the case when no access is possible. The probability of duplication is therefore the same. It is however, reduced when the incumbent has not full bargaining power. In the extreme case where the incumbent has no bargaining power no duplication takes place. The fixed charge is then zero and the resulting contract equivalent to an ex-post linear access contract. Overall, non linear contract types are therefore a useful instrument as they allow separating objectives maximising investment incentives.

When industry demand instead is price dependent an increase in the linear access price leads to higher retail prices but also a decrease of demand for the access seeking firm and the investor. There is therefore no one-to-one pass through anymore creating an asymmetry between the firms as the investor in its hinterland incurs only its marginal network cost and not an (above marginal cost) opportunity costs. The incumbent will therefore charge a lower uniform price than the competitor and have a relatively higher market share in the competitive segment (partial foreclosure). This outcome is therefore different to the outcome an integrated monopolist (joint-venture) would prefer creating allocative inefficiency and reducing overall rent extraction. As a consequence, duplication can now occur as the competitors' profits under duplication may be higher than under access given that lower marginal costs would now allow the entrant to increase its demand, especially in its own hinterland. When the access seeking firm increases its bargaining power, finally, the contracted linear access price will be reduced, leading to lower retail prices of both firms and a relatively higher market share of the access seeker. With non-linear access prices instead, again, higher investment incentives can be achieved. Setting a fixed charge, the incumbent can reduce the variable access fee returning to a more allocatively efficient and symmetric solution, while not being able to reach the joint venture allocation (as the incumbent cannot control the access conditions for both firms)[77]. An optimal allocation without necessitating a joint-venture allowing still full rent extraction could possibly be reached with an even more complex tariff foreseeing next to the fixed fee also the distribution of an adequately chosen ex-post lump sum transfer according market shares (similar to the "compensation mechanism" proposed by some Swiss operators).

---

[76] both operators would then face this access cost instead of the access price and there would be no wholesale market anymore

[77] Changes in the bargaining power would here again not change the allocation.

Finally, when a linear or non-linear binding contract is instead signed ex-ante the competitor may commit to a usage-based and possibly also a fixed charge for access. When ex-ante negotiations break down, the outside profits depend on the outside option scenario (no access or linear or non-linear ex-post NGN access). Under ex-post contracts, as shown, a hold-up problem may arise, where not the entire rent can be extracted from the competitor in case the incumbent does not have full bargaining power. For the incumbent, the investment is then already sunk at the time of negotiation. It will therefore not be considered during an ex-post bargaining stage (e.g. Nash bargaining), the outside option being that only the incumbent offers NGN based products. When ex-ante contracts are used instead, investment costs are not sunk at the time of negotiation and the hold-up problem can be mitigated (and it even disappears with sufficiently complex contracts). Investment cost can therefore be shared somehow with the entrant. The option for an ex-ante contract correspondingly increases the incumbents' profits under price independent demand, (weakly) increasing the range of investment costs that it can sustain and therefore total coverage when compared to the corresponding ex-post contracts. Also, duplication can be avoided, as under this ex-ante contract ex-post the fixed charge is already sunk not creating any incentive for duplication for the competitor. Under price dependent demand, this result does not necessarily hold, as a reduction of the access cost from building own duplicated infrastructure can lead to an increase in the competitors' demand, which may potentially be profit enhancing. In case where a fixed contribution is sunk, this reduces, however, such incentives also in this case. Duplication is therefore in any case more limited under ex-ante contracts. Overall, compared to ex-post contracts (and no access), ex-ante contracts in general provide higher investment incentives (for total coverage) while minimizing duplication and dampening competition if the regulator does not put in place safeguards. This even occurs without considering uncertainty or risk aversion due to bargaining advantages.

b) Uncertainty

Inderst and Peitz (2013) consider a similar model as Inderst and Peitz (2012a), introducing uncertainty about the success of the investment. In addition, the effects of risk aversion and investment timing are analysed. Differently to Inderst and Peitz (2012a), however, duplication is a priori assumed to be not economically feasible facilitating the analysis.

Uncertainty is here introduced by assuming that the NGN gross utility is drawn from a distribution function with values equal to (fail) or higher than the gross utility derived from copper (success to the extent of the utility difference). When both operators use the NGN with respect to the situation where copper is used, an increase in the gross utility of the NGN does affect price and profits only under price dependent demand.

Table 6 summarizes the predicted effects of different access options on investment incentives under uncertainty and risk neutrality, assuming that granting access generates value (net increase in industry profits), i.e. that there is sufficient expansion of total demand and/or lessening of competition so that foreclosure is not an optimal strategy for the incumbent.

	Hold-up problem	Usage of NGN by competitor in all cases	Competitors' outside option	Overall NGN investment incentives
Fixed access charges unconditional on NGN gross utility				
1) - Ex-ante contract - Non-optional fixed charge unconditional on demand	Efficient	No	- Incumbent NGN/copper - Competitor copper	Intermediate
2) - Ex-post contract (before realisation of demand) - Optional fixed charge unconditional on demand	Inefficient	No	- Incumbent NGN - Competitor copper	Low
Fixed access charges conditional on realisation of NGN gross utility				
3) - Ex-post contract (after realisation) - Optional fixed charge conditional on demand	Inefficient	Yes	- Incumbent NGN - Competitor copper	Intermediate (maximum with full bargaining power)
4) - Ex-ante contract - Optional fixed charge conditional on demand	Efficient	Yes	- Incumbent NGN - Competitor copper	Maximum
5) - No fixed charge - Linear usage based charge	Inefficient	Yes	- Incumbent NGN - Competitor copper	Intermediate (but higher than unconditional fixed fee)
6) - No fixed charge - Nonlinear usage based charge	Inefficient	Yes	- Incumbent NGN - Competitor copper	Lower than than linear usage based charges

Table 6 - Effects of different access options on investment incentives under uncertainty and risk neutrality

Under **non-optional fixed fees,** the access seeker enters a binding **ex-ante** agreement on an access charge plan and there is no opt out possibility. It is assumed that after signing the contract a fixed charge (investment contribution) has to be paid by the competitor in any case and usage based access will be granted ex-post at marginal cost (as in all other cases below when a fixed charge is considered). The access seeker is, however, free to buy zero quantity after realization of demand, meaning that only the fixed charge is non-optional. The allocation on the retail market would then be the same as under duplication (symmetric) as both competitors would enjoy marginal costs access ex-post. The fixed contribution can have two effects on coverage. In case the incumbents' investment would be viable also without it (when the competitor would continue to use copper), total coverage is not affected. Access is still granted in this case as long as it creates added value for the industry (extension of total demand and/or lessening of competition). In cases when the investment without the investment contribution of the competitor is not viable, coverage is, instead, extended when compared to no NGN access. The operators will in this case be able to agree on an ex-ante fixed fee as long as industry profits under NGN (both firms) exceed industry profits under copper (both firms) by more than the investment cost (via extension of total demand and/or lessening of competition). Such a scheme does, however, not provide for maximum investment incentives as differently to optional plans described below the outside option for the competitor is in one scenario for the incumbent based on copper reducing the incumbents bargaining power and extractable rents[78].

---

[78] It is also shown as an example that the access option of setting the non-optional ex-ante fixed charge at the investment cost multiplied by the expected market share of the competitor would not necessarily satisfy the participation constraint in the case when a single investment is not profitable but a co-investment is. There may

Under an optional unconditional fixed fee, the competitor has the possibility to seek access signing an access contract ex-post or also ex-ante, while it can then also opt out of the contract after uncertainty has resolved and it is known whether demand is high or low. The competitor will accept to pay the agreed fixed fee in case demand (gross utility under NGN) turns out to be sufficiently high. In this case, in fact, its copper based profits would otherwise be too importantly reduced by customers switching to the incumbents NGN products. Conversely, when demand turns out to be sufficiently low, the competitor will continue to use regulated copper access, which is socially inefficient, reducing competition and not allowing any rent extraction for the incumbent. When demand turns out to be higher than the level to make the competitors' entry via NGN access viable, the competitor makes positive profits, which can, in addition, not be extracted by the incumbent with an unconditioned fee. The investor then receives the fixed contribution with the probability that demand realizes sufficiently high to make the NGN access contract viable for the competitor. If such a probability is low, the investor would have to increase the investment contribution to obtain a given fraction of the investment. But then again the level of demand necessary to sustain such a charge for the competitor increases, reducing the probability of success, and so on. In other terms, it may be impossible for the incumbent to extract sufficient rent to sustain the investment with an unconditional charge. In addition, this scheme could (at least ex-post) not efficiently address a hold-up problem when the incumbent has not full bargaining power.

The shortcomings of optional contracts can be overcome by conditioning the fixed charge on the realization of demand. When negotiations take place ex-post and after realization of demand for instance, the level of demand (NGN gross utility) can be observed and taken into account at the contracting stage, allowing an efficient adaption to market conditions and efficient surplus extraction. When the incumbent has full bargaining power it can extract the entire profits the competitor generates from upgrading to NGN under any realization of demand. NGN access is therefore here always provided as long as industry profits increase with the introduction of NGN as assumed initially. When considering full bargaining power rent extraction and efficiency is enhanced when compared to an ex-ante unconditional access option where in one scenario the outside option is copper not only for the competitor but also for the incumbent. Under conditional (ex-post) contracts, instead, the outside option is always NGN for the incumbent, who has always already invested[79], and copper for the competitor, putting the competitor in a weaker position. The extractable gross profit from access for the competitor is therefore higher under conditional optional contracts.

As shown under certainty in Inderst and Peitz (2012a) with ex-post contracts the investment incentives for the investor are, however, reduced when it does not dispose of full bargaining power. Ex-ante contracts may solve also this hold-up problem. The same is true under uncertainty. Also, ex-post contracts were shown to be an efficient tool to extract rent as they can be fully conditioned on the actual realization of demand. In principle, it is possible to combine both schemes introducing flexible ex-ante contracts depending on demand realization (as long as the level contracted upon is not only observable but also verifiable ex-post). An optimal access option could therefore be an optional ex-ante contract conditioned on realised demand. In such a case, however, from a practical point of view a series of access prices would need to be defined ex-ante for all possible outcomes. Even though the negotiation here takes place ex-ante, the outside option considered is never that of no investment (where both firms use copper), as the situations defined in the ex-ante contract apply only to situations when the investment would have already been undertaken. In this case the same efficiency as with ex-post contracts can be achieved with ex-ante contracts, while addressing in addition a possible hold-up problem. Compared to a non-optional ex-ante fee where in some cases the outside option consists in no investment by the incumbent and therefore relatively higher profits for the competitor when remaining on copper, the rent possibly extracted by the incumbent is therefore increased. When the outcome can be perfectly observed and verified an ex-ante conditional optional fee would therefore

---

therefore be cases where benefits and therefore the investment contribution would need to be distributed differently.

[79] Or is foreseen to have invested.

provide the same investment incentives as an ex-post optional fee under full bargaining power. When instead the incumbent does not have full bargaining power the ex-ante optional conditional fixed charge is the most efficient tool to promote investment incentives, as it also addresses the hold-up problem. As will be seen in the next section, such an access scheme undermines, however, one of the main functions of a co-investment, which is to reduce the investors' risk, as the investor would in this case need to bear a larger share of the investment cost when demand turns out to be low. In this scenario under risk neutrality this effect needs not to be considered.

Inderst and Peitz (2013) also compare linear usage based charges, assuming that the fixed charge is zero. In this case, any access plan is optional as the competitor could always opt-out by buying zero quantity. As shown under certainty, when demand is price dependent, usage based charges introduce inefficient allocative asymmetries. Nevertheless, investment incentives compared to unconditional fixed fees with equivalent wholesale revenues are shown to be usually enhanced as usage based charges provide conditional wholesale revenues by construction. Also, corresponding non-linear usage based access charges can be considered. When still considering an access scheme that implies the same level of wholesale revenues than under the unconditional optional fixed charge and the linear usage based charge, a non-linear charge such as quantity discount leads to relatively lower access prices when demand is high and relatively higher access charges when demand is low. This has two effects. On one hand, this creates an incentive for both firms to increase outputs when they use NGN, reducing deadweight loss and enhancing competition compared to the linear charge. This usually would lead to lower profits and investment incentives though. On the other hand, when demand is realized to be low, access charges increase relatively, meaning that the likelihood that NGN is used by the competitor is reduced and that usage is less efficient. Overall, investment incentives seem to be lower in case of risk neutrality than with a corresponding linear usage based charge. In addition, negative quantity discounts could also be considered. This is for instance the case with capacity limits, where once reached, higher per unit access costs need to be paid. The conclusions are similar to positive quantity discounts. Capacity constraints could therefore be efficient to increase investment incentives. The authors finally consider a combined fixed and usage based charge under uncertainty. They propose a standard case of a non-optional fixed ex-ante fee and an ex-post optional usage based access fee. The usage based fee can as shown under certainty be used to relax competition in the retail market, increasing investment incentives, while the ex-ante non-optional fixed fee may be used to distribute rents especially when the incumbent does not have full bargaining power. However, with respect to the joint-venture outcome in case of price dependent demand, there continues to exist an allocative inefficiency.

Risk averse firms consider profits less valuable when they are uncertain. The two competitors may also have different levels of risk aversion, for instance resulting from their varying ability to access the capital market. Inderst and Peitz (2013) then consider an ex-ante non-optional fixed fee (with the usual marginal cost usage based charge) and alternatively a linear usage based charge (above marginal cost) generating a priori the same wholesale revenues. In this case, when demand turns out to be high, it is shown that the investor has higher total profits under the usage based charge than under the fixed charge. Also, when demand turns out to be low, the investor would have lower profits under the usage based charge. The profit function of the incumbent under a usage based charge is therefore rotated with respect to profits under the fixed charge. The investors' profits with a fixed charge over all possible outcomes of demand are therefore less risky than under usage based charge. The latter therefore shift more risk to the investor. Conversely, the risk the competitor would bear with a non-optional fixed fee would be the same as the investors'. If regulation aims at balancing risks between market participants such an access option could therefore be desirable[80] and depending on the extent of risk adversity of the incumbent this could increase investment incentives accordingly. When considering (unconditional) optional fixed charges instead the risk profiles changes radically. In this case when demand turns out to be low the competitor would opt not to ask for access. From a certain

---

[80] Abstracting from a possible foreclosure or late entrant problem.

level of demand, it would ask for access and pay the fixed fee. The investors profit function is therefore shown to have a discontinuity (increase) at some level of realised NGN gross utility. The level of the discontinuity depends on the level of competition. When there is weak competition (strong horizontal differentiation) the discontinuity corresponds nearly to the fixed charge implying a large revenue risk for the investor.

Finally, Inderst and Peitz (2013) also introduce a dynamic model, where demand for NGN in the market is expected to exogenously grow over time, meaning that operators may prefer waiting some time before investing. Investment can in a basic scenario be seen as an initial decision causing a number of periods of profits depending on the realisations of demand for NGN. Also, from the moment the competitor asks for access, it is supposed to need to pay a corresponding fixed charge also in each following period to access the network. This setting implies that there is an optimal moment for the competitor to invest and adopt NGN via access, the moment being determined by the paths of the access charges and gross profits. Introducing uncertainty about the NGN gross utility means that waiting is becoming an even more attractive option. But, as the NGN already exists, waiting is not socially optimal. Therefore, the fixed charge should be set low initially and rise over time. This could then be an efficient access option for ensuring earliest possible NGN adoption by the competitor while maximising investment incentives. In an additional scenario when the investor is allowed to dilute its investment over time and when cumulative investments are assumed to increases the likelihood of high NGN gross utility realisations, there may – under uncertainty - also be value of waiting for the investor, especially for risk averse investors. Comparing a fixed to a linear usage based fee in this context, it is shown that the latter may have an efficiency advantage over the former as it would increase with the competitors' subscribers over time while fixed revenues would remain constant. For a given level of investment contribution, the usage based fee may, therefore, lead to relatively quicker investments and more efficiency.

3.3.5.   Empirical literature

Empirical data on the effectiveness on new regulatory options such as co-investments is by definition not available. Krämer & Vogelsang (2012) provide, however, a laboratory experiment on the effects of a co-investment option in the market which can be empirically analysed. In their model two firms determine the coverage of their NGN networks in a Greenfield in three areas: metropolitan, urban and rural (respectively increasing in investment costs per household). Depending on the scenario a firm can roll-out independently or (partially) cooperatively. In subsequent ten stages firms compete repeatedly à la Bertrand in a retail market with homogeneous goods in all areas where they have own infrastructure or access (at a geographically uniform price). When the price of two operators is the same, customers are supposed to have a higher probability to choose the incumbent (75%). Access regulation (LRIC[81]) is exogenous and assumed to be in effect wherever only one firm is present. In the scenario without a co-investment option the incumbent first and then the entrant decide on their independent coverage. When instead a co-investments option is admitted, the two firms can, in a prior stage, agree bindingly on the area they will cover by co-investment (basic investment sharing, where the total investment cost for the infrastructure is assumed to remain unchanged). After agreeing on a co-investment, the operators again choose their independent coverage. Under these model settings in the last stage prices would in equilibrium be competed down to marginal costs and the market would be split. In a finitely repeated setting the unique equilibrium of the whole retail game is equivalent. The marginal cost to which prices are competed down includes, however, not only the average marginal cost for access on the other operators network but also the opportunity cost in form of an own (average) access price (represented by the average marginal cost for access for the other operator). This is the case, as giving up a customer implies that the operator does not have to pay an (average) access fee anymore, but that in turn it will receive an (average) access fee. Regarding the investment stage, under the independent investment scheme the authors find that the first mover advantage of

---

[81] Including a return on investment

the incumbent leads to an equilibrium such that it would cover all possibly profitable areas with own infrastructure anticipating that uncovered profitable areas would otherwise be covered by the entrant (in which case its overall profits would decrease as it would have to pay a positive ROI to the entrant for access). It is also found that the entrant having the same cost structure would find it unprofitable to invest in additional areas and that duplication is not feasible as the entrant would need to pay investment costs in own infrastructure without being able to obtain any additional profits (no wholesale profits and retail profits are always zero). In equilibrium, therefore, the incumbent rolls out as far as profitable alone and the entrant asks for access. In the investment stage under the scheme which foresees the possibility for co-investment the equilibrium outcome is surprisingly shown to be identical. As a co-investor, the entrant would have access to the infrastructure at marginal costs not needing to pay any ROI to the incumbent via an access charge. When deciding for co-investment, however, wholesale profits are the only real benefit of investment as retail prices are competed down to marginal cost. Any extent of co-investment would therefore reduce the overall profitability of the infrastructure. Thus, co-investment is fully avoided in equilibrium. After unsuccessful co-investment talks, the equilibrium outcome would then be the same as under independent investment with the incumbent covering all profitable areas and the entrant asking for regulated access.

In a laboratory experiment the authors then tried to evaluate differences between these scenarios. In addition to the scenarios described the participants in the experiment were also exposed to an outside scenario under independent investment with communication where similar to the co-investment scenario they could communicate before the investment stage (but not make a co-investment contract). Such a scheme is unlikely to exist in reality. In both cases, however, participants could not communicate about prices (Chinese wall). In a first empirical model, a mixed-effects linear regression is used to test for differences in total coverage and collusion across different scenarios (Table 7).

	Investment (total coverage)	Intensity of Competition
Co-investment option	Intermediate	Lowest
Independent networks investment option	Intermediate	Intermediate
Independent networks investment option – with limited communication possibility	Maximum	Intermediate

Table 7 – Experimental results, effect of availability of options on competition and investment

In a first econometric analysis, it is found that in an artificial scenario with independent investment, the possibility of communication leads to highest total coverage. The co-investment option scheme leads to less but not statistically significantly different total coverage, when compared to the standard independent investment scenario. Interestingly, even though not an equilibrium outcome under the co-investment option, 56% of duopolies chose to co-invest. This could be motivated by the second result. The authors also use the model to test for differences in the average level of price collusion (over ten periods) in form of a simple Lerner index and a variant of the Lerner index measuring the deviation from average costs. The result shows that collusion is significantly higher in the scenario with a co-investment option present when compared to the other scenarios. Finally, a three level model is estimated considering single periods. These regressions show that tacit collusion decreases towards the end of the game. The authors suspect, however, that this is due to the finite nature of the game. More importantly, it is shown that collusion increases from round to round. Therefore, the longer the participants are in the market, the more they learn to collude.

In a second econometric analysis, the influence of actual market outcomes such as the share of co-investment coverage (rather than differences in scenarios) on total coverage, prices and consumer welfare is estimated. The authors state that they did not impose any demand or cost shocks, meaning that differences in prices or total coverage could be caused only by the conduct of the firms (collusion and investment levels). They assume therefore the absence of any endogeneity problem and use simple regressions where the explanatory variables are treated as exogenous. Such a fully exogenous setting is unlikely to be realistic and results could be unstable. The most important results seem however to broadly support to preceding analysis that the possibility of communication per se significantly increases coverage. Moreover, the share of co-investment coverage (excluding effects related to communication) would not increase total coverage. Regarding collusion it is found that the share of duplication as expected reduces the level of collusion while co-investment increases it (even net of communications effects). The authors see the latter effect as a mystery and speculate about a psychological result from a stronger bond between the two firms in the case of co-investment. Overall they show that consumer welfare can be increased via co-investment when regulators are able to hold these collusive effects somehow in check.

3.4. Conclusion

In this section the conclusions holding throughout the literature and possible future work in this field are described. Directly comparing the results of the theoretical literature is a complex task, as fundamentally different market models and co-investment agreement details are considered. Despite these differences, however, the conclusions and recommendations offered by the literature are surprisingly consistent.

Generally, co-investment agreements are shown to always increase investment incentives in duopoly coverage when compared to no access, while usually not having an impact on total coverage. Total coverage can, however, be affected too with co-investment agreements when compared to the outside option, as they can be used to reduce downstream competition (via internal and/or external access prices, by communication or other means), to extract more rent from access seekers, to extend total demand or in case of risk averse operators to share risks. The fine details of such agreements as well as of the considered outside options therefore matter.

- Cambini and Silvestri (2013) show that under certainty and without outsiders, basic sharing is superior to NGN access regulation at marginal cost in terms of welfare, increasing both investment levels and competition, as the competitors' profits may also be taken into account in the investment decision, thereby expanding network coverage at unchanged access conditions. These results remain valid when outsiders are considered even though co-investment schemes can then lead to foreclosure.

- Under uncertainty, without outsiders, when there is differing ability to increase willingness to pay of consumers across firms, this result remains substantially valid according to Cambini and Silvestri (2012). Basic sharing would still provide maximum output while investment incentives are reduced. When the regulator would set the access price at marginal cost, however, basic sharing would continue to provide also higher investment incentives. When the competitor is slightly better than the incumbent in selling NGN services (a regulator would then set the access price to zero), basic sharing continues overall to be the socially optimal choice. When instead the incumbent is (slightly) better, basic sharing is still a better choice than traditional regulation (but not necessarily than deregulation). Nietsche and Wiethaus (2011) find that with equal ability to increase willingness to pay of consumers across firms in terms of consumer welfare, this conclusion remains valid for different forms of access regulation such as LRIC or FDC.

These different authors seem to agree that basic sharing may represent a valid alternative to traditional access regulation. A basic sharing option could in practice be implemented by imposing regulated conditions to NGA joint-ventures, which includes the imposition of an internal ex-post access fees and the split of investment costs. In substance, this is the regulatory scheme implemented in France. The question then arises, however, whether a solution where ex-post regulated NGN access to the infrastructure is continued in parallel to such regulation would not be an even better solution.

- From the literature only few conclusions can be obtained regarding co-investment schemes under a traditional usage based NGN access regulation environment. Bourreau, Cambini and Hoernig (2013) analyse such a setting, however, and conclude that with uncertainty and outsiders deregulation of basic sharing agreements (i.e. no ex-post regulation of the outsider access price) may be socially preferable to access regulation only when services are highly differentiated and the access charge under regulation would be high. This is the case because with outsiders dampening of competition takes place also under basic sharing. Nevertheless, there are some specific circumstances under which deregulation can be a welfare optimal solution in presence of such a co-investment scheme.

Regulators should therefore consider the possibility of deregulation of co-investments and articulate ex-ante which detailed forms of co-investments would warrant which type of deregulation and under which external circumstances. In light of the above result it seems, however, likely that the introduction of a regulated co-investment option should usually be accompanied by continued traditional NGN regulation to hold excessive negative competitive effects due to the presence of outsiders in check[82]. This is also the approach the French NRA has chosen.

- Regarding long-term access options Inderst and Peitz (2012a) show, under certainty, with price independent demand and full bargaining power that non-linear ex-post access fees can increase rent extraction over linear access prices to the point to reach investment incentives under monopoly (joint-venture). This is the case because under price-independent demand, no allocative inefficiencies from access arise. When instead industry demand is price dependent, there is an inherent allocative inefficiency, implying that under any form of (long term) access, investment incentives are reduced. Under these circumstances, a highly complex contract with lump-sum compensation payments based on ex-post market shares can possibly achieve replication of the monopoly outcome under full bargaining power and certainty. Finally, ex-ante contracts increase investment incentives for any tariff plan when the incumbent does not have full bargaining power, making rent extraction always more efficient.

- Under uncertainty instead, Inderst and Peitz (2013) show that the above is no longer true and that fixed unconditional fees are inefficient as when demand turns out to be low the competitor would continue to use the copper network. Competition as well as investment incentives could, however, be enhanced when it would be given access at reasonable terms. Conditional fees are therefore more efficient in this case. Conditional fees can also be defined ex-ante (describing all possible outcomes), additionally addressing a possible hold-up problem. Ex-ante optional conditional fixed fees are therefore the most efficient (fixed only) access option to promote investment incentives under risk neutrality. Finally, with risk aversion, it is shown that profits are less valuable when they are uncertain. When the investor is known to be risk averse and regulation aims at balancing risks between market participants a largely non-optional ex-ante fee becomes again an interesting access option promoting investments.

- The empirical literature is still limited. In Krämer & Vogelsang (2012) co-investment is not taking place in equilibrium due to unrealistically aggressive downstream retail competition assumptions when compared to the rest of the literature. Unsurprisingly, their experimental results suggest that such an equilibrium would not arise in reality and that operators may use co-investments as a means to increase collusion - even when the internal access fee is fixed at marginal cost and in presence of Chinese walls limiting communication.

---

[82] It should be noted here that Cambini and Silvestri (2013) show that when considering basic sharing as an alternative to traditional regulation with outsiders, basic sharing would be preferable for regulators to access regulation (at marginal cost) even though this may imply foreclosure.

To conclude, also on the subject of co-investments many issues still remain to be explored. The most important flaw when comparing theoretical literature with applied regulation seems that multifibre has not yet received attention in academic research. Given the attention this roll-out option has received from regulators as well as Governments and the European Commission, future co-investment models should try to incorporate multifibre options. The main properties of multifibre, which could allow integration in existing models, are that it allows more flexibility and independence via IRUs when compared to traditional networks, that it enables consumers to purchase services from multiple providers simultaneously and that switching costs are reduced. More concretely, multifibre may allow physical infrastructure competition between the partners. In the existing literature usually under joint-ventures a common access price to the infrastructure for outsiders is chosen by the partners jointly and under long-term access an incumbent is setting this price (alone). With multifibre instead both types of outsider access charges could be set independently by the two partners. In addition, another form of access debated by regulators has not yet received attention. Participation in a co-investment agreement could also be possible ex-post. Such a scenario seems particularly relevant in the multifibre case, where for instance in Switzerland two dedicated fibres (out of four) are today usually left unused. Also, the co-investment compensation mechanism described in the section on regulatory practice has been only broadly explored by Inderst and Peitz (2013). It should be analysed in more detail in a fully fledged model. Finally, there is yet no common framework to date that allows for instance a direct comparison of the Dutch co-investment case (joint-venture) to the predominant Swiss co-investment case (long-term access agreements).

4. Concluding remarks

This chapter integrates themes which have appeared throughout the text.

- The review of practical cases has shown that by the end of 2013, European regulators continued to lack clarity on how to handle co-investment agreements and geographic regulation. At the time of writing, a wide variety of regulations were being applied. Their ultimate success will not be evident until several years after their implementation. To cite only the most extreme cases which have been reviewed:

    - While nearly all regulatory authorities continue to apply uniform access prices, the Dutch regulator imposes regional access prices varying with the extent of investment cost.
    - Regarding co-investment, on one hand, the Swiss regulator leaves full freedom to co-investors to shape their NGN risk sharing agreements (as long as compatible with cartel law). On the other hand, the French regulator regulates all important clauses of such agreements (share of investment cost to bear, access price for insiders and outsiders, location of distribution point).

- To date, there do not appear to be strong initiatives to address these issues at the European level. It is possible that this is the case as regulators, BEREC and the European Commission do not yet have a clear vision on these issues. This is understandable, to some extent, as for example the analysed effects of co-investment schemes depend on the fine details of such agreements and often also on market parameters such as the willingness to pay, investment requirements or potential industry demand expansion. While the economic literature on these topics is still limited, it seems, however, to clearly show that co-investment agreements with the right clauses can enhance welfare over traditional regulation at least in some cases and that exclusively uniform usage based access pricing may no longer be optimal. Future literature will likely further build on this and provide more stable insights. Nevertheless, it seems that regulators are now in a position to start to reflect on how to introduce and implement regional access prices and to promote co-investments.

Annex

Table 8 - Theoretical analyses – geographic segmentation of remedies and geographic aspects of regulation

	Main Assumptions							Remedies considered	Case of free wholesale market	Main results
	Geographic difference in cost and competition considered	Geographically differentiated retail prices allowed	Allow for geographically differentiated access prices (cost)	Allow for geographically differentiated access prices (competition)	Type of retail competition	Entry	Presence of old technology			
Bourreau, Cambini, & Hoernig (2012b)	Yes	Yes	Yes	Yes, number of firms	Bertrand, horizontally differentiated good	Two potential incumbents and potential downstream entrants	No	-Access price regulation -Access obligation	Yes. Bertrand, no differentiation (at same prices access providing firm is chosen randomly)	Cost-based geographic access prices lead to suboptimal roll-out and duplication and uniform pricing to too much duplication. The paper analyzes geographic regulatory instruments able to achieve the social optimum, e.g. geographically differentiated prices or remedies.
De Matos & Ferreira (2011)	Yes	Yes	Yes	No (assumed to be competitive when investment costs are such to allow infrastructure competition)	Cournot, horizontally differentiated good	Endogenous (simulation)	No	Access price regulation	No	Different market outcomes with different access rates are simulated. Low access prices erode profitability of infrastructure providers. When regional markets interact, deregulation of more competitive areas may trigger a monopoly situation in an adjacent market.
Flacher & Jennequin (2012)	Yes	No	No	No	Cournot, vertically and horizontally differentiated good	One potential infrastructure entrant, one potential downstream entrant (no duplication)	Yes	-Access price regulation -Coverage obligation	Yes	Show that regulation for maximize total coverage (full deregulation) is not optimal, as well as cost-based regulation to maximize static efficiency. Suggests that setting access prices and coverage obligations is optimal.
Lestage & Flacher (2010)	Yes	Yes	No	No	Bertrand, vertically differentiated good	Two potential incumbents	Yes	- Access price regulation	Yes	A low access price may lead to areas having two equilibria, where it is not clear which operator would invest. It is then uncertain whether there will be investment. If the quality advantage of firm A is sufficient this problem disappears.

Table 9 – Theoretical analyses – NGN co-investments

Cooperation type	Paper	Main assumptions						Main results
		Fixed investment contribution (share of investment cost)	Usage based access charges for insiders	Usage based access charges for outsiders	Uncertainty	Presence of old technology	Effect of NGN	
Joint-venture (JV)	Cambini & Silvestri (2013)	Yes, equal shares	Yes (free choice)	Yes, positive and higher than insider fee	No	Yes	NGN increases willingness to pay (same for both firms) depending on investment extent	Cambini and Silvestri (2013) show that without outsiders, basic sharing is superior to NGN access regulation at marginal cost in terms of welfare, increasing both investment levels and competition, as the competitors profits may also be taken into account in the investment decision, thereby expanding network coverage at unchanged access conditions. These results remain valid when outsiders are considered even though co-investment schemes can then lead to foreclosure.
	Cambini & Silvestri (2012)	Yes, variable shares.	Yes (free choice)	-	Yes	Yes	Chance that NGN investment increases willingness to pay (by same amount for both firms)	Under uncertainty, without outsiders, when there is differing ability to increase willingness to pay of consumers across firms basic sharing always leads to more competition and output than with regulation or deregulation, while full deregulation induces the highest investments. From a welfare point of view, when the competitor is better than the incumbent in providing NGN services (and the regulator would consequently set the NGN access price under full regulation to zero) basic sharing is always optimal. When instead the incumbent is better, the ranking is less clear. Basic sharing usually continues to be optimal.
	Cambini & Silvestri (2013)	(see above)	Yes, marginal cost	(see above)	(see above)	(see above)	(see above)	(see above)
	Nietsche & Wiethaus (2011)	Yes, equal shares	Yes, marginal cost	-	Yes	Yes	Chance that NGN investment increases willingness to pay (by same amount for both firms)	Risk sharing (basic sharing) is shown to lead to maximum output and competition as well as to maximum consumer welfare, when compared to LRIC, FDC or deregulation, for its strong competitive effects and reasonable investment incentives allowing the operators to share benefits and costs upfront - even if ex-post the investment fails.
Basic investment sharing (particular form of JV)	Bourreau, Cambini & Hoernig (2013)	Yes, equal shares	Yes, marginal cost	Yes, same as insider fee	Yes	No	Demand for NGN can be high or low (same willingness to pay across firms)	With uncertainty and outsiders deregulation of basic sharing agreements (i.e. no ex-post regulation of the outsider access price) may be socially preferable to access regulation only when services are highly differentiated and the access charge under regulation would be high. This is the case because with outsiders dampening of competition takes place also under basic sharing. Nevertheless, there are some circumstances under which deregulation can be a welfare optimal solution in presence of such a co-investment scheme.
	Krämer & Vogelsang (2012)	Yes, 75% incumbent / 50% competitor (according to demand share)	Yes, marginal cost	-	No	No	No quality effect, willingness to pay is identical for both firms	Basic sharing is not taking place in equilibrium due to aggressive downstream retail competition assumptions when compared to the rest of the literature. Experimental results suggest that such equilibrium would not arise in reality and that operators may use co-investments here as a means to increase collusion - even when the access fee is fixed at marginal cost and in presence of Chinese walls limiting communication.

Overall the regulator can ensure positive effects on consumer welfare when the introduction of a co-investment option is accompanied by measures preventing collusion.

Access innovation on joint-venture	Mizuno (2009)	Yes, variable	Incumbent has access at marginal cost. Competitor has access at regulated prices (fixed multiple of marginal cost)	-	No	No	NGN investments have no effect on quality but can reduce marginal costs	Under a regulated (usage) cost based access pricing rule when positive spill-overs from access innovation on the entrant (via a high access charge) are sufficiently high, the entrant also benefits from a reduction in access costs. In this case the negative effects from competition (in this range the incumbents marginal costs decrease more than the entrants') are sufficiently balanced. Then the entrant may participate in a cooperative investment scheme increasing overall investment incentives. The author moreover shows that in case of standard LRIC cooperation is enhancing total welfare. Finally he shows that investment incentives under no cooperation can be enhanced with a two-part tariff but that this would not be welfare optimal.
	Inderst & Peitz (2012a)	-	Incumbent has access at marginal cost. Competitor has access at possibly above marginal oost prices.	-	No	Yes	NGN increases consumers' gross utility of the service (same amount for both operators).	Under certainty, with price independent demand and full bargaining power that non-linear ex-post access fees can increase rent extraction over linear access prices to the point to reach investment incentives under monopoly (joint-venture). This is the case because under price-independent demand, no allocative inefficiencies from access arise. When instead industry demand is price dependent, there is an inherent allocative inefficiency, implying that under any form of (long term) access, investment incentives are reduced. Under these circumstances, a highly complex contract with lump-sum compensation payments based on ex-post market shares can possibly achieve replication of the monopoly outcome under full bargaining power and certainty. Finally, ex-ante contracts increase investment incentives for any tariff plan when the incumbent does not have full bargaining power, making rent extraction always more efficient.
Long term access	Inderst & Peitz (2013)	-	Incumbent has access at marginal cost. Competitor has access at possibly different access options.	-	Yes	Yes	NGN increases consumers' gross utility of the service (same amount for both operators).	Under uncertainty instead conclusions of Inderst and Peitz (2012a) are no longer true and fixed unconditional fees are inefficient. When demand turns out to be low the competitor would continue to use the copper network. Competition as well as investment incentives could, however, be enhanced when it would be given access at reasonable terms. Conditional fees are therefore more efficient in this case. Conditional fees can also be defined ex-ante (describing all possible outcomes), additionally addressing a possible hold-up problem. Ex-ante optional conditional fixed fees (with subsequent access at marginal cost) are therefore the most efficient access option to promote investment incentives under risk neutrality. Finally, with risk aversion, it is shown that profits are less valuable when they are uncertain. When the investor is known to be risk averse and regulation aims at balancing risks between market participants a largely non-optional ex-ante fee becomes again an interesting access option promoting investments.

Bibliography

Akamai. 2012. The state of the internet.

Anacom. (2009). Markets for the Supply of Wholesale (Physical) Network Infrastructure Access at a Fixed Location and Wholesale Broadband Access.

Anton, J.J., Vander Weide, J.H., & Vettas, N. (1999). Strategic Pricing and Entry under Universal Service and Cross-Market Price Constraints: Duke University, Department of Economics.

ARCEP. (2009). Recommandation de l'Autorité de régulation des communications électroniques et des postes relative aux modalités de l'accès aux lignes de communications électroniques à très haut débit en fibre optique.

Autoriteit Consument en Markt. (2013). Tariefbesluit ontbundelde glastoegang (FTTH) 2012. Decision DTVP/2013/201158.

Avenali, A., Matteucci, G., & Reverberi, P. (2010). Dynamic access pricing and investment in alternative infrastructures. *International Journal of Industrial Organization, 28*(2), 167-175.

BEREC. (2012a). Report on Co-investment and SMP in NGA networks. *(12)41.*

BEREC. (2012b). Report on the Implementation of the NGA Recommendation. *(11)43.*

BEREC. (2010a). Next Generation Access – Implementation Issues and Wholesale Products. *(10)08.*

BEREC. (2010b). Report on self-supply.*(10)09.*

BEREC. (2013). Common Position on Geographic Aspects of Market Analysis. Public consultation document. *(13)186.*

Bourreau, M., Cambini, C., & Doğan, P. (2012). Access pricing, competition, and incentives to migrate from "old" to "new" technology. *International Journal of Industrial Organization.*

Bourreau, M., Cambini, C., & Hoernig, S. (2010). National FTTH plans in france, italy and portugal. *Communications and Strategies*(78), 107-126.

Bourreau, M., Cambini, C., & Hoernig, S. (2012a). Ex ante regulation and co-investment in the transition to next generation access. *Telecommunications Policy, 36*(5), 399-406.

Bourreau, M., Cambini, C., & Hoernig, S. (2012b). Geographic Access Rules and Investment. *CEPR-Centre for Economic Policy Research, Discussion Paper Series, 9013,* 1-48.

Bourreau, M., Cambini, C., & Hoernig, S. (2013). Cooperative Investment, Uncertainty and Access. *CEPR-Centre for Economic Policy Research, Discussion Paper Series, 9376,* 1-36.

Bourreau, M., & Doğan, P. (2005). Unbundling the local loop. *European Economic Review, 49*(1), 173-199.

Bresnahan, T.F., & Salop, S.C. (1986). Quantifying the competitive effects of production joint ventures. *International Journal of Industrial Organization, 4*(2), 155-175.

Cambini, C., & Jiang, Y. (2009). Broadband investment and regulation: A literature review. *Telecommunications Policy, 33*(10), 559-574.

Cambini, C., & Silvestri, V. (2012). Technology investment and alternative regulatory regimes with demand uncertainty. *Information Economics and Policy, 24,* 212-230.

Cambini, C., & Silvestri, V. (2013). Investment Sharing in Broadband Networks. *Working Paper.*

Cave, M. (2008). Building the broadband network. In Australia's Broadband Future: Four doors to greater competition. *Committee for the Economic Development of Australia (CEDA), Growth No. 60*

Choné, P., Flochel, L., & Perrot, A. (2000). Universal service obligations and competition. *Information Economics and Policy, 12*(3), 249-259.

Choné, P., Flochel, L., & Perrot, A. (2002). Allocating and funding universal service obligations in a competitive market. *International Journal of Industrial Organization, 20*(9), 1247-1276.

De Bijl, P., & Peitz, M. (2006). Local loop unbundling: One-way access and imperfect competition. In: R. Dewenter, & J. Haucap (Eds.). Access pricing: Theory and practice (91–117). Vol. 86. Elsevier.

de Matos, M. G., & Ferreira, P. (2011). *Entry in Telecommunications' Markets.* Mimeo.

de Streel, A. (2010). Market Definition in the Electronic Communication Sector. *Telecommunications, Broadcasting and the Internet: EU Competition Law and Regulation, 3rd ed., Sweet & Maxwell,* 411-435.

Elixmann, D., Ilic, D., Neumann, K.-H., & Plückebaum, T. (2008). The Economics of Next Generation Access. *WIK-Consult Report for the European Competitive Telecommunication Association (ECTA).*

European Commission. (2002). Guidelines on market analysis and the assessment of significant market power under the Community regulatory framework for electronic communications networks and services.

European Commission. (2010a). Recommendation on regulated access to next generation access networks (NGA). Official Journal of the European Communities, Bruxelles.

European Commission. (2010b). A digital Agenda for Europe. Official Journal of the European Communities, Bruxelles. COM(2010) 245 final/2. Retrieved from http://eur-lex.europa.eu/LexUriServ/LexUriServ.do?uri=COM:2010:0245:FIN:EN:PDF

European Commission (2011). Digital agenda scoreboard 2013. Retrieved from https://ec.europa.eu/digital-agenda/

European Commission (2013). Proposal for a Regulation of the European Parliament and of the Council on measures to reduce the cost of deploying high-speed electronic communications networks. COM(2013) 147 final.

Flacher, D., & Jennequin, H. (2012). Access regulation and geographic deployment of a new generation infrastructure. Mimeo.

Foros, Ø. (2004). Strategic investments with spillovers, vertical integration and foreclosure in the broadband access market. *International Journal of Industrial Organization, 22*(1), 1-24.

Foros, Ø., & Kind, H.J. (2003). The broadband access market: Competition, uniform pricing and geographical coverage. *Journal of Regulatory Economics, 23*(3), 215-235.

Fudenberg, D., & Tirole, J. (1984). The fat-cat effect, the puppy-dog ploy, and the lean and hungry look. *The American Economic Review, 74*(2), 361-366.

Hoernig, S.H. (2006). Should uniform pricing constraints be imposed on entrants? *Journal of Regulatory Economics, 30*(2), 199-216.

Hoernig, S., Jay, S., Neu, W., Neumann, K. H., Plückebaum, T., & Vogelsang, I. (2012). Wholesale pricing, NGA take-up and competition. *Communications and Strategies*, (86), 153.

Hoernig, S., Jay, S., Neumann, K.-H., Peitz, M., Plückebaum, T., & Vogelsang, I. (2012). The impact of different fibre access network technologies on cost, competition and welfare. *Telecommunications Policy, 36*(2), 96-112.

Houpis, G., Santamaria, J., & Lucena Betriu, J. (2011). Geographic segmentation of broadband markets: appropriate differentiation or risk to a single EU market? *Communications and Strategies*(82), 105-126.

Ilic, D., Neumann, K., Plückebaum, T. (2009). Szenarien einer nationalen Glasfaserausbaustrategie in der Schweiz. *WIK-Consult Report*.

Inderst, R., J. Kühling, K.-H. Neumann and M. Peitz (2012), Investitionen, Wettbewerb und Netzzugang bei NGA

Inderst, R., & Peitz, M. (2012a). Network investment, access and competition. Telecommunications Policy, 36(5), 407-418.

Inderst, R., & Peitz, M. (2012b). Market asymmetries and investments in next generation access networks. *Review of Network Economics, 11*(1).

Inderst, R., & Peitz, M.(2013). Investment Under Uncertainty and Regulation of New Access Networks. *ZEW-Centre for European Economic Research Discussion Paper*, (13-020).

Informa. 2011. Romania: Where the incumbent is not the incumbent.

Jay, S., Neumann, K.-H., & Plückebaum, T. (2011). Critical market shares for investors and access seekers and competitive models in fibre networks. *WIK-Consult Report*.

Katz, M.L., & Shapiro, C. (1985). Network externalities, competition, and compatibility. *The American economic review, 75*(3), 424-440.

Katz, R.L., Vaterlaus, S., Zenhäusern, P., & Suter, S. (2010). Polynomics. The impact of broadband on jobs and the German economy. *Intereconomics, 45*(1), 26-34.

Krämer, J., & Vogelsang, I. (2012a). Co-investments and tacit collusion in regulated network industries: Experimental evidence. *Available at SSRN 2119927*.

Laffont, J.J., & Tirole, J. (2001). *Competition in telecommunications*: the MIT Press.

Lestage, R., & Flacher, D. (2010). Telecommunications infrastructure investment: Access regulation and geographical coverage: Mimeo. Retrieved from http://www.webmeets.com/files/papers/EARIE/2010/73/Lestage-Flacher%2C2010%20%28EARIE%29.pdf

Manenti, F., & Scialà, A. (2011). Access Regulation, Entry, and Investment in Telecommunications.

Middleton, C., & Van Gorp, A. (2010). Fiber to the home unbundling and retail competition: developments in the Netherlands. *Communications and Strategies*,78(2), 87-106.

Mizuno, K. (2009). Comparison of investment regimes with cost-based access pricing rules. *Japan and the World Economy, 21*(3), 248-255.

Mölleryd, B.G. (2011). Network sharing and co-investments in NGN as a way to fulfill the goal with the digital agenda.

Neu, W., Neumann, K. H., & Vogelsang, I. (2012). Analyse von alternativen Methoden zur Preisregulierung.Analyse von alternativen Methoden zur Preisregulierung. Studie für das Bundesamt für Kommunikation. *WIK-Consult Report.*

Nitsche, R. (2010). NGA: Access regulation, investment and welfare. A model based comparative analysis.

Nitsche, R., & Wiethaus, L. (2011). Access regulation and investment in next generation networks—A ranking of regulatory regimes. *International Journal of Industrial Organization, 29*(2), 263-272.

OECD. (2010). Geographically segmented regulation for telecommunications, DSTI/ICCP/CISP(2009)6/FINAL

OPTA. (2008). Policy rules. Tariff regulation for unbundled fibre access. Decision OPTA/AM/2008/202874

Pereira, J. P., & Ferreira, P. (2011, January). Next Generation Access Networks (NGANs) and the geographical segmentation of markets. In *ICN 2011, The Tenth International Conference on Networks* (pp. 69-74).

Plum (2011). Costing methodology and the transition to next generation access. Report for ETNO.

Polynomics. (2009). Fibre Platform Competition, New York, Olten, 10. Juli 2009.

Riordan, M. H. (1992). Regulation and preemptive technology adoption. *The Rand Journal of Economics*, 334-349.

Röller, L.-H., & Waverman, L. (2001). Telecommunications infrastructure and economic development: A simultaneous approach. *American Economic Review*, 909-923.

Schneir, J. R., & Xiong, Y. (2012). Strategic and economic aspects of network sharing in FTTH/PON architectures. In *23rd European Regional ITS Conference, Vienna 2012* (No. 60397). International Telecommunications Society (ITS).

Shubik, M., & Levitan, R. (1980). *Market structure and behavior:* Harvard University Press Cambridge.

Stockdale, D. (2011). Geographically Segmented Regulation: Lessons from the FCC for European Communications Markets. *Communications and Strategies*(82), 85-104.

Valletti, T.M., Hoernig, S., & Barros, P.P. (2002). Universal service and entry: The role of uniform pricing and coverage constraints. *Journal of Regulatory Economics, 21*(2), 169-190.

van Dijk. (2012). Broadband Internet Access Cost. Report for the European Commission. September 2012.

Xavier, P., & Ypsilanti, D. (2011). Geographically segmented regulation for telecommunications: lessons from experience. *info, 13*(2), 3-18.

Zenhäusern, P., Suter, S., & Vaterlaus, S. (2010). Plattformwettbewerb und regulatorische Empfehlungen. Polynomics Studie.

Chapter II

Entry and Competition in Local Newspaper Retail Markets

When two are enough

This chapter estimates sustainable coverage and competitive effects of entry for Swiss newspaper sellers which sell composite goods, including a range of other products such as food and near-food items. It utilises the applied entry threshold ratio methodology from Bresnahan and Reiss (1991), which allows estimation even when the range of products under examination is not exactly defined and when price and quantity data are not available. It is found that under monopoly pricing, single firm entry is sustainable in communes with a market size of over 482 people (leaving 310 Swiss communes without a selling point). With duopoly prices, instead, a first firm would only be able to enter a market with 921 people (leaving another 263 communes, corresponding to 2,1% of the population, without coverage). There are therefore tangible benefits from above duopoly prices in monopoly regions. Thus, a clear and quantifiable trade-off between prices in monopoly regions and coverage exists. Moreover, it is found that a second entrant in this market strongly increases competition, while further entry doesn't have significant additional competitive effects. From a welfare perspective, therefore, "two is enough" to ensure competition in this market. It is shown that this is not the case in some other retail markets, where entry by a third firm may significantly affect competition. Finally, using the estimation results, it is show that the public policy, which consisted of having the government controlled Swiss Post enter the newspaper sellers' retail market, was not optimal as it was focused on urban areas where neither coverage nor competition could be enhanced, while risking competitive distortions. At the same time, it is shown that there are communes in which such a government policy may be welfare enhancing.

1. Introduction

This paper estimates sustainable coverage and competitive effects of entry for Swiss newspaper sellers which sell composite goods, including a range of other products such as food and near-food items. It is found that when communes are not too small, the presence of one or more newspaper selling points can be sustained by the free market. Moreover, the effect of a second entrant on competition is shown to be strong, while subsequent entry does not have further significant impact on competition. This result is important, as there was repeated direct public intervention in local newspaper sellers' markets (by the Swiss Post) which are shown to be competitive. Moreover, it is shown that the market dynamics on the distribution market may also affect upstream competition between newspaper publishers, as, for instance, in the absence of local sales channels entry barriers are increased.

The media is often called the fourth[1] of Montesquieu's three powers of the state and a founding pillar of modern democracy (Montesquieu (1751)). People like Alain Peyrefitte and Thomas Jefferson argued that a society needs freedom of speech and the freedom of press[2] to ensure checks and balances between the people and the three traditional powers. In Switzerland, for instance, freedom of the press seems to be granted in general by the legal, political and economic framework. Freedom House ranks Switzerland respectively fourth, third and fifth in the world (Freedom House, 2012). In this setting legal freedom of the press alone is, however, not a sufficient condition for efficient public information. It is recognised that the press in general has the ability to influence the public opinion. This can occur through a topic selection bias, e.g. by omitting relevant information or by publishing biased information. Sufficient competition between journals might provide for optimal horizontal differentiation (Salop, 1979), leading to a homogeneous coverage of all topics "requested" by readers – and eliminating to a large extent a possible topic selection bias or biased opinions in the industry as a whole. Bignon and Miscio (2010) show that competition between newspapers produces coverage of all relevant information in the market. (Blasco & Sobbrio, 2012) show that a sufficiently high degree of competition in the market for newspaper drives out possible media biases also in relation to particular interests of advertisers and review relevant literature. Freedom House notes that in practice, despite a good overall ranking, the situation in Switzerland is not unproblematic as large publishing houses control most of the print sector and that concentration of ownership has forced many stand-alone newspapers to merge or shut down. A lack of competition implying a local informational bias might therefore potentially represent a democratic challenge. Such an issue might be particularly relevant in Swiss regions (cantons) and communes where political decisions are regularly put at vote[3].

In addition to freedom of the press and competition between publishers, accessibility also matters. The public should be able to access information without unreasonable effort. In rural areas, for example, if there are no newspaper selling points in the commune, it may be difficult for a citizen not subscribed to a newspaper to find a newspaper to buy nearby. This article analyses under which circumstances entry and therefore coverage is viable. It is concluded that only in 310 from 2'700 communes (corresponding to 1%

---

[1] The idea of the Fourth Power is not explicitly found in Montesquieu. It appears during the first part of the 19th Century in writings Thomas Carlyle (Carlyle, 1993) citing British politician Edmund Burke (1729-1797). Carlyle and others as Macaulay coined the expression Fourth Estate to describe an informational power, which included the political reporters of the Press, but also, in a larger sense, all writers and the entire institution of literature. In modern times the fourth branch or power has a meaning which includes other mass media such as TV and radio.

[2] Jefferson (1787): "The basis of our government being the opinion of the people, the very first object should be to keep that right; and were it left to me to decide whether we should have a government without newspapers, or newspapers without a government, I should not hesitate a moment to prefer the latter."

[3] More concretely, a citizen of the commune of Zürich needs to be able to participate at votes on four dates per year on 25 law proposals often very different in their nature and at a general election. This requires a large and broad amount of local information in order to be able to make sufficiently informed choices.

of Swiss population) no selling point is economically viable. Moreover, the impact of entry on monopoly, duopoly and oligopoly newspaper seller markets in Swiss communes is assessed. An entry threshold model based on Bresnahan and Reiss (1991) is used to estimate minimum demand levels necessary for entry of a specific number of firms as well as the impact of entry on newspaper sellers average variable profits. Most importantly, these estimates show that in communes with only one newspaper seller, there are considerable margins, while margins are strongly reduced when a competitor enters (considering their full product portfolio). Entry by subsequent competitors is shown to have no significant further effects on margins. It is shown that this is a particular feature of the Swiss newspaper sellers' market and that in different other retail markets entry of a third firm also has substantial competitive effects. While the high profitability of monopoly selling points may be seen by some as a problem, it is argued that it has also a positive effect, increasing the geographical coverage of newspaper selling points. More concretely, it is shown that when a monopolist would be able to charge only duopoly prices, coverage would be lost in 263 communes (2,1% of the population). There are therefore tangible benefits from above duopoly prices and a clear and quantifiable trade-off between prices in monopoly regions and coverage exists. For a detailed theoretical model regarding the interplay of local competition, investment and coverage with two firms and increasing fixed costs towards rural areas, the reader may refer to Bourreau, Cambini and Hoernig (2012). Unlike newspapers, the availability of other media like TV, Radio or the Internet is usually largely independent from geographically located intermediaries. Nevertheless, coverage matters, especially for broadband Internet where coverage is not available in some rural areas. Such other forms of media will be shown to not represent valid substitutes to newspapers (not sufficient in-depth information) to date. In addition, local radio and TV stations are often not available in rural areas. The impact of other forms of media is therefore largely abstracted from in the model.

It is additionally argued in this article that the relatively strong margin earned by monopoly newspaper selling points over the whole range of products sold, may not only be beneficial to accessibility, but may also correspond to increased competition between newspaper publishers. Regarding this second effect, Shapiro and Varian (1999) show that newspapers (and information services in general) are *experience goods*; consumer must experience them to value them. To be able to efficiently compete, firms must therefore offer consumers a way to easily experience the good before committing to a subscription. The easiest way to experience a newspaper is usually to buy a single copy at a selling point. While newspaper selling point sales may well have a lower weight than subscription sales when considering total sales of publishers, they may play a crucial role during entry for acquisition of customers, and be a prerequisite for effective competition between newspapers. If all newspaper selling point sales channels in an area were to disappear, this would clearly increase experience costs (and entry barriers) and therefore correspond to less upstream competition between editors. Concretely, for instance, it might be very difficult for a new small local newspaper to gain a customer base in rural area, where nearly no selling points are available. Paradoxically, high (overall) retail profitability of newspaper selling points can therefore increase total coverage (accessibility) and thereby competition between editors[4] (reducing informational bias) - potentially benefiting the public in two distinct ways. These benefits do, however, not come without a cost, which may be mainly related to relatively higher prices of the whole portfolio of goods sold by newspaper sellers in areas with limited competition.

This paper consists of a series of chapters laying down a theoretical model, adjusting it to the case under analysis and analysing econometric results. Chapter 2 describes the market environment in newspaper

---

[4] It should be noted here that competition at wholesale level is not generally independent of retail competition. Inderst and Valletti (2007) show how retail competition affects wholesale competition via indirect effects. In practice an upstream supplier may be unable to raise (upstream) prices substantially above competitive levels as this would make its retailers less competitive and customers may switch to retailers of the competitor. This may be the case when there are different wholesale suppliers or a vertically integrated firm. Here, as will be shown, there is a publishing-wholesale distribution monopoly supplying inputs to all retailers. In such case there are no indirect effects.

distribution as well as publishing markets in Switzerland. In Chapter 3, the current public policy in place will be described. Chapter 4 describes the theoretical model used in this paper and Chapter 5 its practical econometric adaption to the Swiss newspaper sellers market. Subsequently, Chapter 6 describes the data used. Chapter 7 summarizes regression results and as well as minimum demand levels necessary for entry (entry thresholds). In addition, entry threshold ratios, indicating the effect of entry on profitability are calculated. Finally, Chapter 8 makes policy recommendation based on these results and Chapter 9 concludes.

2. Market environment

The newspaper distribution market is part of a complex system of up- and downstream markets (Figure 1). Generally, there is a two-sided newspaper market where readers buy newspapers and read them and an advertising market were advertisers buy advertising space in these newspapers (higher readership is typically associated with higher prices). Newspapers are then distributed in two different ways: one is distribution through the subscription channel where copies are sent to a home address of the customer and distribution through newspaper selling points, where usually there are wholesalers and retailers involved. These markets and their relationships are described in the following sections.

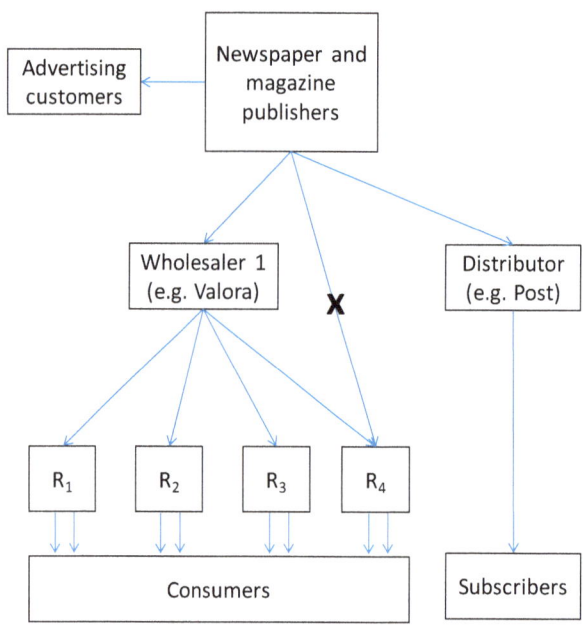

Figure 1 - Dependencies between the players in the newspaper value chain (OFT, 2009)

On the reader market side, publishers compete to supply newspapers to consumers by charging a cover price for print copies (whether distributed singly, over the counter, or by subscription). On the other side of the market, both free and non-free newspapers compete for advertising revenues.

Possible media substitutes

In general, competition authorities have regarded TV, radio and Internet content as distinct markets from printed press mostly because there is a different range and depth of information (e.g. European

Commission in Recoletos / Unedisa or UK Competition Commission, in the Newsquest Ltd Independent News and Media plc). In particular, for the OFT (2009) it is often possible for newspapers to cover more of the background to a news story than can be covered in TV or radio news. The U.S. Department of Justice (2011) has a similar view. It states that it had concluded in past investigations that non-newspaper media do not sufficiently constrain the pricing of single newspapers, advertisements, the pricing of newspaper subscriptions or newspapers' investments in news and editorial content and thus are not in the same market. It does, however, not exclude future changes in this definition[5]. In Switzerland, the Competition Appeal Authority broadly shares these views and sees radio, TV and the Internet as complements to the newspaper readers markets rather than as substitutes (Berner Zeitung AG/Tamedia AG/Wettbewerbskommission[6] and Tamedia AG/Espace Media). In addition, it does not consider mobile online news as a substitute to commuters dailies (accessed and read while commuting to work, e.g. in the train), among other things because of the small size of mobile phone displays.

In an early study, Ahlers (2006) shows that, in practice, there was a certain migration of traditional reading to online reading, but that the magnitude was limited. The study shows that at that time only 12% of U.S. adults were Internet news only users. This confronts with 22% of multichannel (on- and offline news readers) and 43% of offline only news readers. However, this is confirmed by a MIS study for Switzerland (M.I.S. Trend, 2009), which states that only 22% of users often use their Internet connection to read the press news (26% from time to time). This is again confirmed by a recent study by Latzer et al. (2012) which states that only 6% of Internet users do not read print newspapers at all and that 83% of them regularly read print newspapers[7]. More than by the content or the niches chosen by online or printed news the choice may be largely motivated also by habit and accessibility (Van der Wurff (2011)). The available information today therefore seems to indicate that, overall, it is not unreasonable to consider the printed press in Switzerland as still a separate market from radio, TV and Internet today and when viewing historical data from the beginning of the 21th Century.

Different types of press products and local potential coverage

Newspaper publishers produce a wide range of press products. The Swiss Competition Commission defines a market for free commuters' dailies independent from the market for other daily newspapers. Such commuters' dailies are usually characterized by very short news without providing further analysis. Commuter's dailies are distributed for free at central nodes of the public transportation system (also using distribution boxes instead of newspaper selling points). A central feature of such press products is their free and timely accessibility in trains when people are commuting. Such accessibility, it concludes, is not given to a sufficient degree by other media such as radio, TV and the Internet (although this may change in the future for example with e-readers). Regarding other newspapers, the Commission thinks that commuters' dailies may not satisfy the needs of a typical reader of a traditional newspaper and that they are therefore to be largely considered as complements. It also states that very few readers seem to exclusively read commuters' dailies. The market entry of the largest commuters' daily "20 minutes", is stated to have had a significant negative impact on single copy sales (dropping around 25% in seven years)[8], but that in the more important subscriptions market (from a revenue point of view) such a

---

[5] The U.S. Department of Justice further states that "this conclusion is perfectly consistent with the observation that newspapers have been losing subscription and advertising revenues to other media, as some degree of competition across market boundaries is the norm. Whether changes in technology and consumer preferences may lead to the conclusion that a relevant market should include sales of advertisements (or content) by both newspapers and other media remains something that should be analyzed on a case-by-case basis."
[6] RPW 2006/2, p.375
[7] Note that 7% of total users indicate that they have cancelled a newspaper subscription because the information was also available on the Internet.
[8] Berner Zeitung AG/20 Minuten

significant reduction could not be verified[9]. For this reason their overall substitutability with traditional newspapers seems to be limited. While the Competition Commission did not make a binding decision in this respect, the appeal body has corrected this by subsequently deciding that there would be no sufficient substitutability with other newspapers and that the reduction of single copy sales may also have other reasons[10]. Regarding magazines, the OFT affirms that on the demand side, there is a wide range of magazine on different subjects and that it would seem unlikely that consumers would substitute, for any given title, another title that had very different content. That would suggest magazines are forming a number of product markets based on their content, target audience and frequency of publication (OFT, 2009). The Swiss Competition Commission has taken a similar position in its proceedings when affirming that there is a separate readers market for different types of magazines[11]. More importantly, it also defines a series of related markets in this context: 1) daily newspapers, which include local, regional, national and international news, 2) a market for newspaper advertising at national level 3) a market for newspaper advertising at regional and local level and finally 4) analogous separate advertising markets on local/regional as well as national level for radio and for TV advertising[12]. The Commission then even proceeds to define the specific geographic boundaries of each local/regional newspaper and advertisement market based mainly on their local reach. Some markets were even defined to be smaller than cantons. Nevertheless, it is found that there are even in urban regions high concentrations in daily (local) newspapers markets[13].

There are various studies analysing competition between local newspapers (Dewenter, 2003). In general the conclusion is that as a consequence of economies of scale in advertising and news production, in many cases, rural areas simply cannot support more than one newspaper publisher with regional coverage implying high market power. Empirical estimations of competition between newspapers are rare, given the complex nature of newspapers as a platform for readers and advertisers. There is literature that is arguing that in some cases, regions may overlap, leading to higher competition in overlap areas (Dertouzos & Trautman, 1990). This seems, however, as shown by the Swiss Competition Commission, to be rarely the case in Switzerland. Even on a national level competitive problems seem to arise (Argentesi & Filistrucchi, 2007; Argentesi & Ivaldi, 2005). The problem of market power of local newspapers is – at least in rural areas – in Switzerland also widely recognized.

From an institutional point of view, this paper focuses on the market for daily newspapers including local and regional content (as well as national and international news). This is the only market allowing for in depth local news coverage in many rural areas, as often no (sufficiently) local radio or TV station is present.

Single copy sales channel (retail level)

In general, two channels for the distribution of newspapers are distinguished: subscriptions and single copy sales at newspaper selling points. In the distribution and retail business operators offer "distribution channels" to distribute their content directly or via intermediaries to final customers at a given price.

---

[9] Berner Zeitung AG/20 Minuten, RPW 2004/2, p. 540
[10] Berner Zeitung AG, Tamedia AG/Wettbewerbskommission in RPW 2006/2, p.347
[11] Tamedia AG/Espace Media, RPW 2007/4 P. 605-629
[12] The differentiation of advertising markets between newspaper and other media seems common as target groups and marketing strategies are different. It can be noted here that the U.S. Federal Communications Commission had shown that cross-price elasticity between retail ads in local newspapers and radio is very small, between local newspapers and TV they even seem to be complementary (Bush, 2002).
[13] RPW 2007/4 P. 605-629

An early proceeding by the Swiss Competition Commission has analysed the market for newspaper selling points in Swiss train stations[14]. It states that in theory single copy sale of newspaper may be possible in the form of a small additional business for a large part of businesses such as Convenience stores (selling food and near-food items[15]), tobacconists, etc. This possibility would, however, be restricted by the fact that wholesalers might ask for minimum revenues. Moreover in other markets restricted search costs a customer is willing to face are defined. For example the Commission had concluded that the market radius for small supermarkets to cover "basic shopping needs" is 10 minutes of travel time[16]. It seems clear that the radius for buying a newspaper would be lower. Similarly, the Office of Fair Trading states that the geographic market is likely to be highly localized, in particular for newspapers, given the need for convenient local purchases (OFT (2009)). There is therefore a very local market for individual newspaper selling points. However, in the model proposed in the next chapters, it is assumed that the geographic markets extend over the area of a commune, when considering only small communes. Finally, while from a supply perspective the subscriptions channel may be a valid substitute for sales to the single copy sales channel, from the demand side there are significant differences. Single copy sales target spontaneous buyers while subscriptions target habitual readers. For this reason the, Competition Commission does not see the two channels as belonging to the same market. In the model proposed, the newspaper sellers' market will therefore be considered in isolation.

Moreover, it should be noted here that single copy sales are particularly important for newly entering local newspaper seeking new customers. Shapiro and Varian (1999) show that newspapers (and information services in general) are *experience goods*, i.e. a consumer must experience them to value them (see also Nelson (1970))[17]. Experience goods have usually lower price elasticity as consumers fear that lower prices may be due to quality issues or unobservable problems. To be able to efficiently compete, firms must therefore offer consumers a way to easily experience the good. The marketing response to this phenomenon has often been to offer consumers free samples of such goods so that they can be aware of their value and (potentially) buy. The easiest way to experience a newspaper is usually to buy a single copy at a selling point. Often, publishers also offer trial subscriptions. In this case, however, this usually involves signing a cancellable contract for a period of one month or more as well as the exchange of address and billing details. If all selling point sales channels in an area were to (locally) disappear, this would clearly increase experience costs and entry barriers and therefore correspond to less upstream competition between publishers. Concretely, for instance it might be very difficult for a new small local newspaper to gain a customer base in rural areas when nearly no selling points were available. Paradoxically, therefore, in this market high (overall) retail profitability of newspaper selling points can increase total coverage (accessibility) and thereby competition between editors upstream (reducing informational bias).

The market regime in practice is described by the Competitions Commission as foreseeing that distributors both at wholesale and at retail level would only be intermediaries (between the publishers and the consumers) receiving each commissions for selling newspapers. Unsold copies are returned to the publisher at no cost[18]. In the Swiss market the direct largest customer of exclusive wholesaler Valora (see also next section) except Valora itself with 20% was Volg, a mainly small rural retailer, with about 11% of selling points supplied. Analysing in detail the data at disposal[19] (Figure 2), newspaper sellers seem to nearly always contemporarily sell other goods, most prominently a more or less broad range of food and near-food items (kiosks, groceries and petrol station make up for about 85% of newspaper selling points).

---

[14] RPW 1999/3, p.403-422
[15] Any other goods of daily use
[16] Valora/Cevanova, RPW 2009/1, p.77-80
[17] While in practice the informational content of every edition of a newspaper is different, it is assumed that once tried consumers can nevertheless value the quality of a particular brand of newspaper.
[18] RPW 1999/3
[19] Source: Presstalis 2008

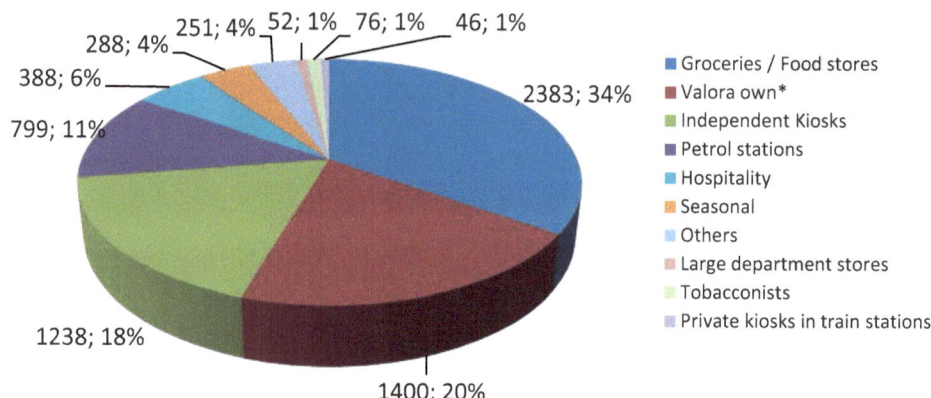

Figure 2 – Number and type of newspaper selling points supplied by Valora, 2008

Next to Valora and Volg, Coop stores with about 10% (retailer) and the Swiss Post (government control) with about 7% of selling points held considerable buyer shares in 2008. Post offices have for a long time distributed newspapers. This aspect will be mentioned in later chapters.

Single copy sales channel (wholesale)

A wholesaler of press products is an intermediary between the publisher producing newspapers and the retailers selling these products to the customers. A retailer is able to substitute a wholesaler only if the alternative wholesaler is able to supply all titles requested. Valora is acting as single exclusive wholesaler of (a large range of) press products of German langue titles[20] in Switzerland. In addition to its wholesale business, Valora is, as shown before, controlling also a number of retail newspaper selling points (mostly kiosks).

An important feature of the demand of newspapers of retailers is that they have to arrive soon in the morning as after 9 a.m. the units sold strongly decline. The Competition Commission has in the indicated merger proceeding defined consequently a wholesale market for the timely (early morning) supply of press products to retail newspaper selling points. It assumes that there are substantive economies of scale and scope in the logistic process constituting properties of a natural monopoly. It states that there would be a potential risk of discrimination in favor of Valora's own selling points or even of foreclosure and that it would intervene under in case such practices would take place. This means that in Switzerland, newspaper selling points should face homogeneous input products when compared to Valora (which does not exclude for example volume discounts a priori). It is also stated that such wholesalers may not have a similar dominant position when facing the publishers. These may have considerable buying power as they might potentially sign a contract with a new outside firm for wholesale distribution (e.g. Post).

Subscription channel

For newspaper subscriptions, publishers seem to need to ensure delivery at home before 6.30 a.m. as this seems to be, according to the Competition Commission, a requirement of the consumers[21]. Normal post delivery is usually completed before 12 a.m. in Switzerland. Nevertheless, the Swiss Post is able to ensure early delivery in most of Switzerland, but it has entered this business only very late (2007). According to the Commission the market shows properties of a natural (regional) monopoly, as the lowest costs are generated when early delivery is ensured by one single firm – similarly to the rest of postal

---

[20] Exclusivity contracts are signed, RPW 1999/3, p.403-422
[21] Post/Tamedia/NZZ

services. In some (few) regions there are, however, still parallel early delivery operators present today, controlled mainly by local newspapers (sometimes in joint-venture) or the Swiss Post[22]. The Commission was critical towards the operators controlled by local newspapers as they could exclude new local newspapers. This conclusion has led to the blocking of a local joint-venture between the Swiss Post and local newspapers.

3. Public policy

The distribution markets under observation described earlier are the single copy as well as the subscription channel distribution of daily newspapers, which include local and regional content as well as national and international news

Single copy sales channel (retail level)

The Post, which is still under full government control in Switzerland, has directly entered the newspaper retail market at different times. It has sold newspapers to final customers in 1'200 post offices from 1993 to 1998. Then, after a period of seven years, it had again entered this market in 2005 with 33 newspaper selling points in its post offices (urban and rural areas)[23] offering a small range of press products, including the most important regional and national daily newspapers and popular magazines (less than 25 titles in total). During a pilot phase, the Post did not work with Valora. Instead, it was directly supplied by the publishers. As the Post ensures early delivery of subscriptions for many local newspapers, such an agreement with publishers seemed possible (at least for a certain amount of time). Valora[24] later declared, however, that this agreement would not respect the long (exclusive) collaboration with the publishers and that it would engage in discussions with them. Subsequently direct negotiations between the publishers and the Swiss Post seemed to break down and the Post communicated after the trial in 2007 that it would "from now on" work with Valora. In 2008, the Post was operating 464 selling points (about 20% of post offices) supplied by Valora. Soon, however, Post seems to have entirely abandoned this business. In 2012 the post offices concerned seem to sell mainly other products such as stationery items, mobile phones and traveler items.

Subscription channel

The new postal law (Postgesetz, art.16)[25] foresees that newspapers which include local and regional content are awarded a reduction in delivery costs of 30 million CHF. Furthermore, 20 million CHF are foreseen for magazines of associations and foundations[26]. The delivery price reduction is valid, however, only for services under universal service (i.e. services from the Swiss Post). This applies therefore not to early delivery (before 12 a.m.), which is a liberalized market (BAKOM, 2012a). A large part of subscriptions are therefore not incentivized. To date the price reduction has therefore been mainly used by weeklies and a few local newspapers (BAKOM, 2012b). Also, wholesale distributors of newspapers to selling points could not profit from these reductions. The main motivation for the above provisions is the

---

[22] This does not necessarily mean that similar competitive conditions are given for all competitors as the Post for example does not need to respect the ban of heavy road transport (over 3,5 tons) on Sundays and during the night - the time early delivery has to be organized.

[23] http://www.post.ch/post-startseite/post-konzern/post-medien/post-archive/2005/post-mm05-pilotversuch-zeitungsverkauf/post-medienmitteilungen.htm

[24] In an interview still in Valora CEO Wüst in 2005

[25] The development of this law over the years is well explained in Ecoplan (2010)

[26] No such reduction is awarded to newspapers with a circulation of more than 100'000 copies. The government can add further criteria to award the reductions such as coverage, periodicity, extent of informational part or extent of advertising part.

strong process of concentration of independent local newspapers in Switzerland (from 45 in the year 2000 to 32 in 2012). The Competition Commission has shown that in many areas there are strong concentrations with local newspaper accounting for often more than 60% of the local market share even when taking into account national titles[27].

To conclude, the newspaper distribution markets under consideration were affected in the past by the entry of public entities in the retail distribution market as well as in the wholesale distribution market for subscriptions. In addition there are subsidies for late delivery, which are used, however, only by a small fraction of newspapers. The following chapter will analyse the level of competition in the newspaper sellers' market and the appropriateness of entry of a public entity in this market.

4. Analytical Framework

Newspapers are generally sold to generate customer frequency ("foot traffic") at selling points, meaning that people may want to look for a place where to buy a newspaper and once there, they would also acquire other products (e.g. food and near-food items, tobacco etc.) (NAA (2012)). In other terms, it is held that selling newspapers is having a positive externality, which a newspaper seller would typically take into account. Cases would therefore even be possible where the seller has an unprofitable standalone business for selling newspapers, but it is reasonable to continue to sell when considering the positive externality on other products and services sold. From this point of view a newspaper selling point satisfies some of the criteria laid down by the Swiss Competition Commission to qualify for a two-sided market (namely positive externality), but as there are not two different customer groups involved, the property cited can be considered as a simple case of complement[28]. In addition, as described earlier, products which are typically sold together with newspapers may be many and different in their nature. Analysing entry and competition based on a structural model estimating demand and supply using prices and quantities of such products would therefore prove impossible. Recent estimation techniques exist, however, which allow empirical measures of competition through the observation of sellers' entry decisions and general demand and supply shifters. This means that it is possible to make inferences on the competitive interactions in a market of newspaper sellers considering a composite good sold without the need to collect data on individual products supposed to be in the affected market. In such case any complementary or substitutive effects between goods sold are internalised.

Based on Chamberlin (1933) and Panzar and Rosse's (1987) theoretical description of free-entry competition, Bresnahan and Reiss (1991) propose an "entry threshold" model which allows estimates of the minimum demand necessary for a specific number of firms to enter the market. Entrants may in this model face entry barriers. What is observed from the data is only whether entry occurs or not. The entry threshold is then defined as the market size required to sustain the entry of a particular number of firms. Following Bresnahan and Reiss (1991) it will be shown how entry thresholds ratios provide a scale free measure of the effect of entry on market power. The objective is to relate shifts in market demand to changes in the equilibrium number of firms. The degree of competitiveness in this market under progressive entry will then be used to evaluate public policy relating to newspaper distribution in Switzerland.

Suppose that d(Z,P) is the demand function of a representative customer, depending on the price of the good (P) and some other demand shifting variables Z (e.g. income). Suppose further that S(Y) is the number of customers, determined by some variables Y (e.g. population). Then total market demand can be expressed by:

---

[27] Tamedia/Espace Media, RPW 2007/4, p.621
[28] Again, as explained by Ordover (2007) this is very similar to analyzing a "two-sided market".

$$Q^d = d(Z, P)S(Y) \tag{1}$$

Further fixed costs depending on some exogenous variables W (e.g. real estate availability in the commune) may be defined:

$$F(W) \tag{2}$$

For newspapers sellers, typical cost functions of local retailers may apply (as for instance for tire dealers in Bresnahan and Reiss (1991)). The standard model assumes that a firm's average and marginal costs are U-shaped. Hence, there are economies of scale at first, and after a certain quantity of composite goods sold, diseconomies of scale. Thus, there is a minimum efficient scale (MES) of output. It should be noted, that the marginal cost functions related to newspapers alone may (slightly) differ. This is assumed, however, to not have a substantial effect on the model, as it can be assumed that newspapers only account for a small part of total sales and margins of a typical newspaper seller. Still, it should be noted that for the sale of newspapers alone, marginal costs are decreasing or possibly constant as wholesale monopolist Valora seems to award quantity discounts[29] – a form of second degree price discrimination - after some levels of output per distribution point are reached (identical for all sellers)[30].

Consequently, with q being firm output, define

$$MC(q, W) \tag{3}$$

$$AVC(q, W) \tag{4}$$

It is assumed for simplification that the goods produced are homogeneous products[31]. It is therefore assumed that a typical newspaper seller would sell units of a virtual, composite good composed of newspapers and other products which are sold together with newspapers (typically some range of food and near-food products[32]). P is the corresponding price of such a composite good. There is assumed to be price flexibility over the whole range of goods sold. It should be noted, however, that such flexibility may not exist for the sale of newspapers alone as newspaper prices are printed on the front-page in Switzerland and are also usually charged. Given that newspapers only represent a small part of overall sales of newspaper sellers this aspect can again be neglected. As this papers' objective is to consider all sellers of newspapers, independently of which other products they sell, the model proposed provides an ideal (and probably the only possible) framework for empirical analysis of market power.

For graphical analysis purposes it can be supposed that each seller has in the long run the same cost functions. Figure 3 illustrates the entry threshold model[33].

---

[29] Concretely, Valora pays commissions per copy sold increasing with quantity. The fact that retailers do not pay for inputs, but are paid by wholesalers for their distribution services, is purely financial and does not change the analysis.
[30] Given the earlier cited decision of the Competition Commission it is likely that the Commission would intervene in case of discriminatory prices.
[31] This is a simplifying assumption, as composite goods sold may have some level of differentiation between sellers.
[32] Both categories are fast moving consumer goods (FMCG)
[33] Note that this graph differs from traditional market demand analysis as the X axis is reporting the per firm quantities and not the market quantities.

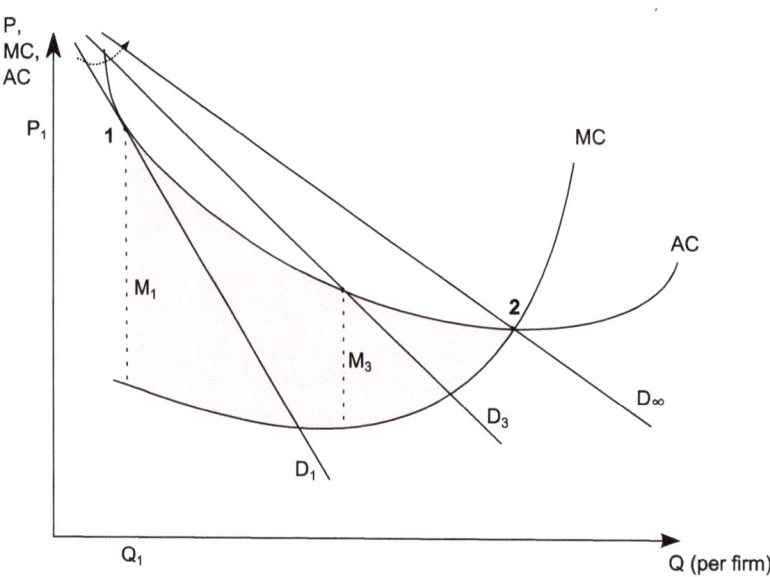

Figure 3 – Entry threshold model (Bresnahan and Reiss (1991))

It will now be analysed what happens if market demand rotates, i.e. if market size $S(Y)$ increases. In the figure per firm demand level D_1 represents the minimal level of per firm demand a single newspaper seller would need to break even and to enter the market ($P=AC$, at point 1). From the level of demand D_1 onwards, a first firm could therefore cover its fixed costs and - potentially - enter the market. Although a newly entered monopolist just breaks even at price P_1, it earns the substantial price-cost margin M_1 for an additional unit (the differential between price and marginal cost). Suppose now that the market size $S(Y)$ further increases. Given that $S(Y)$ is multiplicative in the total demand function (where individual demand is supposed to be independent of the size of the market), total demand continues to rotate outwards. At a given price, the monopoly seller would therefore face more demand and become more and more profitable (total profit). However, the effect of rotating demand is twofold: It is not only increasing actual profits of the firm(s) in the market but it also increases the profits of potential entrants (it is assumed that in case of entry the market is split and all operator obtain the same proportional share of total demand).

As market demand continues to rotate outwards – and so per firm demand (Figure 3) – it will eventually reach level D_2 (or more generally D_N): the minimum per firm demand a potential second (or more generally N-th) entrant would need to enter and break even. When an additional firm enters it would be expected that the new more competitive situation would lead to declining individual margins. Therefore, the more demand continues to rotate outwards, the more firms will enter and the more individual margins decline (at entry level).

When market demand rotates even further outwards, per firm demand grows ever larger in relation to minimum efficient scale (*MES*) and allows for more and more firms to enter. Eventually, per firm demand (D_∞) will reach *MES*, where it will jointly with *MC* intersect *AC* and where margins, therefore, are competitive (zero). At this point, total demand may well continue to increase to infinity, but per firm demand would not be able to increase anymore, as a given firm which is already in the market would have zero profits and would make negative profits when increasing scale above *MES*. A new firm would therefore enter and produce, as all firms already in the market, at *MES*. D_∞ therefore represents the maximum per firm demand.

Given that profitability is expected to decline with each additional selling point (at entry level) it must be expected that entrants will require more per firm demand to enter the later they enter. In particular a duopolist would need a higher number of customers to enter the market than a monopolist, which unlike the duopolist would profit from more favourable competitive circumstances (monopoly prices).

To conclude, in Figure 3 it can be observed that at the level of demand D_1, S_1 consumers pay price P_1 and the monopolist earns a price cost margin of P_1-MC_1. This is just enough to cover its fixed costs at this level of demand (minimal demand level of entry). When market demand grows further, and per firm demand rotates outwards, individual firms' margins decline at entry level as the number of firms in the market increases (from M_1 to M_3 to 0 at D_∞). Eventually, a competitive outcome at *MES* is observed, where margins are zero, and, being at *MES*, total profits are also zero.

This analysis has considered the whole range of goods sold by newspaper sellers (composite good). When considering the sale of newspapers alone, the above assumptions may not be given. When moving, for instance, from a one to a two firm setting, at entry level the composite output per firm increases. Thereby, newspaper quantity discounts might also increase. At entry levels, there may therefore be a reduction in marginal costs related to newspapers alone with entry. At the same time, newspaper sellers do not have pricing flexibility and cannot adjust their prices in response to successive entry and competition. For the sale of newspapers alone margins at entry levels may therefore paradoxically increase (or remain constant) with entry (at entry level). This is, however, assumed not to have a substantial effect on the model, as newspapers only account for a small part of sales and margins of newspaper sellers.

Formalisation

The objective of the model is to relate the number of firms in the market *N* to the size of the market *S* at entry level. Therefore, as the size of the market increases (i.e. demand rotates outward), how does it affect entry? The consequences of entry by relating shifts (respectively rotations) in market demand to changes in the equilibrium number of firms can be examined. The basic concept which will be used to build the model is the one illustrated in Figure 3: the zero-profit conditions indicate the "entry level" of demand for a monopoly, a duopoly, or in general a n-oligopoly firm.

A monopolist's break-even-condition is that profits (for a given size of the market) are zero. The size of the market which satisfies this equation gives the trigger market size for a monopolist to (potentially) enter the market (*S1*). Formally,

$$\Pi_1(S_1) = [P_1 - AVC(q_1, W)]d(Z, P_1)\frac{S_1}{1} - F(W) = 0 \tag{5}$$

Where *AVC* is average variable cost. This expression is necessary as fixed costs need to be specified separately from variable costs in this model.

$$[P_1 - AVC(q_1, W)]d(Z, P_1)$$

is therefore the variable profit per customer (equal to the variable profit per unit times the number of units demanded per consumer). Now solve for the entry market size or "entry threshold" where a monopolist first breaks even:

$$S_1 = \frac{F(W)}{[P_1 - AVC(q_1, W)]d(Z, P_1)/1} \tag{6}$$

This is the market size (number of customers) needed for one firm (the first) to enter the market. The higher the fixed costs or the lower the variable profits per customer, the higher the minimal market size required for one firm, the monopolist, to enter.

Now, assume the market size (and per firm demand) grows and firms enter as in the graphical illustration. Then, Equation 5 can be generalized to describe the zero profit function of an N-th operator to enter the market, assuming that the market size is split equally among the N firms in the market:

$$\Pi_N(S_N) = \left[P_N - AVC(q_N, W)\right]\frac{d(Z, P_N)S_N}{N} - F(W) = 0 \tag{7}$$

This can be represented graphically by Figure 4.

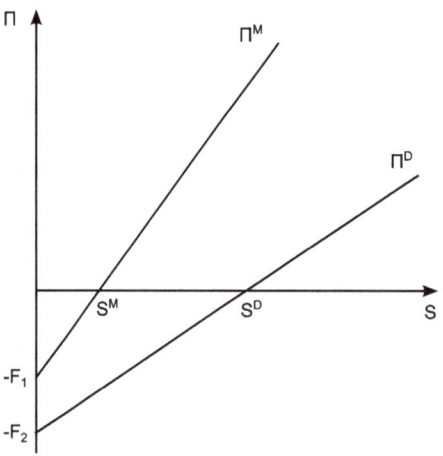

Figure 4 – Entry thresholds for a monopoly and a duopoly

The slopes of the profit functions are the variable profits per customer in the market in the cases of monopoly and duopoly. Naturally, a monopolist reaches the break-even-point earlier than a duopolist as demand increases. The market entry threshold for a second, duopoly firm, S^D is always bigger than two times S^M (firm entry threshold ratio of at least unity). The reason for this is that the average profitability of a firm in duopoly is lower due to competition or, also, because the fixed costs of the second firm are higher (opportunities for instance are rare, e.g. buying adequate real estate for selling space etc.) or most probably because of a combination of these two factors.

Now the *"entry threshold"*, i.e. the minimal total market size necessary for N identical firms to enter (S_N) is:

$$S_N = \frac{N F(W)}{\left[P_N - AVC(q_N, W)\right]d(Z, P_N)} \tag{8}$$

The *"per firm entry threshold" (or entry level of demand)* s_N is the minimal market size that is needed for each single one of the N firms to enter, i.e.

$$s_N = \frac{S_N}{N} \tag{9}$$

$$s_N = \frac{F(W)}{[P_N - AVC(q_N,W)]d(Z,P_N)} \tag{10}$$

With increasing market demand this threshold converges to some value, i.e.

$$s_\infty = \lim_{N \longrightarrow \infty} \frac{S_N}{N}$$

where s_∞ is the per firm demand able to sustain any number of firms (*MES*). Finally, the *"per firm entry threshold ratio"* may be defined as follows

$$s_{N+1}/s_N = \frac{F_{N+1}(W)}{F_N(W)} \frac{[P_N - AVC(q_N,W)]d(Z,P_N)}{[P_{N+1} - AVC(q_{N+1},W)]d(Z,P_{N+1})} \tag{11}$$

The per firm entry threshold ratios, which are estimated in this paper, measure the fall in variable profits per unit by going from a market with *N* firms to a market with *N+1* firms when holding fixed costs and demand per consumer constant. Accordingly, this is a useful expression indicating how much variable profits decline as competition becomes more intense due to entry. It should be noted that this is no measure of the absolute level of profits or competition but on their development when moving from a market with *N* firms to a market with *N+1* firms.

If:

a) Firms have the same fixed costs and

b) Competitive conduct does not change (constant (total) variable profits) with entry

Then, the per firm entry threshold ratio from moving from an industry with *N* players to an industry with *N+1* players is

$$s_{N+1}/s_N = 1 \tag{12}$$

When, instead, ratios higher than one are measured, this must mean that average profitability drops with entry and that a new entrant would need more per firm demand than the preceding entrant to break even (still assuming constant fixed costs) to absorb the increase in competition. Departures of entry thresholds from one therefore measure how competitive conduct changes with successive entry.

Conversely, in an already competitive equilibrium, if firms have the same fixed costs and entry does not change competitive conduct (profitability) anymore it can be expected that the ratio remains one. In this case it would take the same additional market size for a firm to enter in the market with *N* firms as in a market with *N+1* firms. As this measure cannot distinguish the levels of competition though, a per firm entry threshold ratio of one is compatible with both perfect competition as well as a perfect cartel. However, if due to entry a strong increase in competition is observed, it can only be concluded that in the situation preceding entry there was significant market power. The entry thresholds - for the mentioned reasons - are a useful scale free measure of competition.

5. Econometric model

The model can be transformed to allow econometric estimation of entry threshold ratios. Variable profits per customer in a market with N firms (V_N) can be defined as:

$$V_N = [P_N - AVC(q_N, W)]d(Z, P_N)$$

Therefore,

$$s_{N+1} / s_N = \frac{V_N}{V_{N+1}} \frac{F_{N+1}(W)}{F_N(W)}$$

Market size S is supposed to be determined by a series of demand rotating variables Y

$$S(Y; \lambda) \tag{13}$$

Where λ is a parameter vector. Suppose now that average variable profitability V_N in a market with N firms decreases with the number of firms in the market and increases with demand shifting variables (e.g. income per capita). The following additive setting will be used, where α parameters are expected to be positive and decreasing with the number of firms, i.e. α_N should be smaller the bigger N.

$$V_N(Z, W, \alpha, \beta) = \alpha_1 + X\beta - \sum_{n=2}^{N} \alpha_n \tag{14}$$

Z represents demand shifters and W (average variable) costs shifters, which are both included in X. Suppose further that fixed costs F are allowed to be increasing with the number of firms (Equation 15).

$$F_N(W, \gamma) = \gamma_1 + \gamma_R W + \sum_{n=2}^{N} \gamma_n \tag{15}$$

This would account for the fact that later entrants may have higher fixed costs, for example because opportunities are rare. Therefore, it can be expected that γ's are positive, but decrease with N. Finally, fixed cost shifters W are included (e.g. real estate prices or availability). For estimation purposes, total profits in a market with N firms can be expressed as

$$\Pi_N = \overline{\Pi}_N + \varepsilon = S(Y, \lambda)V_N(Z, W, \alpha, \beta) - F_N(W, \gamma) + \varepsilon \tag{16}$$

where $\overline{\Pi}_N$ is the latent variable and ε, the error term, is assumed to be the profit which cannot be observed. This setting allows for a probit estimation in the next section. The error term is assumed to follow a normal distribution, to be independent and to be identically distributed across the markets. Moreover, the term is assumed to have mean zero and variance one throughout. The estimation framework described in the next chapter will allow for prediction of the number of firms in the market.

Estimation procedure

The entry decision of a firm depends on whether an additional firm can earn positive profits. But also the entry decision of the firm that had entered previously to the new entrant will need to be taken into account. If this firm is in the market, it must mean that it earns positive profits. Formally, if profits of the first firm are negative, no entry is possible. If such profits are positive, a firm can enter. Profits in monopoly, however, have to be lower than those necessary to sustain two firms, as otherwise two firms would enter. Therefore, when considering only three states (zero, one or two firms):

$$\Pr(N=0) = \Pr(\Pi_1 < 0) = \Pr(\overline{\Pi}_1 + \varepsilon < 0) = \Pr(\varepsilon < -\overline{\Pi}_1) = \Pr(\varepsilon > \overline{\Pi}_1) = 1 - \Pr(\varepsilon \le \overline{\Pi}_1) = 1 - \Phi(\overline{\Pi}_1)$$

$$\Pr(N=1) = \Pr(\overline{\Pi}_1 + \varepsilon \ge 0 \, and \, \overline{\Pi}_2 + \varepsilon < 0) = \Pr(\varepsilon \ge -\overline{\Pi}_1 \, and \, \varepsilon < -\overline{\Pi}_2) = \Pr(\varepsilon < \overline{\Pi}_1 \, and \, \varepsilon \ge \overline{\Pi}_2)$$

$$= \Phi(\overline{\Pi}_1) - \Phi(\overline{\Pi}_2)$$

$$\Pr(N=2) = \Pr(\overline{\Pi}_2 + \varepsilon \ge 0) = \Pr(\varepsilon \ge -\overline{\Pi}_2) = \Pr(\varepsilon < \overline{\Pi}_2) = \Phi(\overline{\Pi}_2) \qquad (17)$$

The above formalised model explains entry by the level of profit in the market. It can, however, not be estimated by a standard regression which would treat the difference between category 0 and 1 in the same way as the difference between category 1 and 2, whereas in fact these differences correspond only to a ranking. The model must therefore be estimated by a model capable to handle discrete ordered choice dependent variables (Maddala, 1983). The model refers to the cumulative normal distribution function (associated to the probit regression). Ordered probit is a common specification for ordinal response models. The model is estimated using standard maximum likelihood procedure. The maximum likelihood can be written as follows

$$L = \prod_{m=1}^{1042} 1(n_m = 0)\left[\Phi(\overline{\Pi}_1))\right] \prod_{m=1}^{1042} 1(n_m = 1)\left[\Phi(\overline{\Pi}_2) - \Phi(\overline{\Pi}_1)\right] \prod_{m=1}^{1042} 1(n_m = 2)\left[1 - \Phi(\overline{\Pi}_2)\right] \qquad (18)$$

where $\Phi()$ is the inverse normal cumulative function, when, as assumed, the single probabilities are independent. The likelihood of the observed outcome is therefore maximised by maximising the product of probabilities for each state. $1(n_m = j)$ is equal to 1 if in market m a number of j firms is found and zero otherwise. Maximising a product is equivalent to maximising the logs of it, i.e.

$$\ln L = \sum_{i=1}^{1042} 1(n_m = 0)\ln(\Phi(S(Y,\lambda)V_1(Z,W,\alpha,\beta) - F_1(W,\gamma)))$$

$$+ \sum_{i=1}^{1042} 1(n_m = 1)\ln\left[\Phi(S(Y,\lambda)V_2(Z,W,\alpha,\beta) - F_2(W,\gamma)) - \Phi(S(Y,\lambda)V_1(Z,W,\alpha,\beta) - F_1(W,\gamma))\right] \qquad (19)$$

$$+ \sum_{i=1}^{1042} 1(n_m = 2)\ln\left[1 - \Phi(S(Y,\lambda)V_2(Z,W,\alpha,\beta) - F_2(W,\gamma))\right]$$

This function can now be maximised with respect to the parameters. The particular challenge of the econometric model with respect to usual ordinal response models is that it is nonlinear.

6. Data on Swiss communes

When defining from a geographical point of view local newspaper selling markets in Switzerland, it may be reasonable to define it to extend to the communal territory (as in Bresnahan and Reiss (1991)). In many cases, in rural areas, a large part of the commune can be reached within a reasonable walking distance. Between two neighbouring rural communes, there is typically a more or less extended area of relatively low population density, which may constitute an obstacle to cross-commune demand. Observations can therefore be considered sufficiently independent, especially in rural areas which constitute the areas of greatest interest. The data summarized in Table 1 is used for estimation.

Variable	Name in dataset	Unit	Definition	Source	N	Mean (abs)	Std. dev. (abs)
N	blick		Number of newspaper selling points (Blick) in each commune, 2004	Ringer/ Blick	1667	3.83	15.7641

Market size

Variable	Name in dataset	Unit	Definition	Source	N	Mean (abs)	Std. dev. (abs)
y_1	pop	000s	Permanent residential population, 2002	BFS	1667	3.11	10.89
y_2	CommutersIN	000s	Number of commuters into the commune, 2004	BFS	1623	1.13	6.82

Representative customer demand shifters

Variable	Name in dataset	Unit	Definition	Source	N	Mean (abs)	Std. dev. (abs)
z_1	inc	kCHF	Average taxable income per commune, 1998	ESTV	1646	60.78	10.9
z_2	old	000s	Residential population having over 65 years per commune, 2004	BFS	1666	0.15	0.04
z_3	foreigners	000s	Foreign residential population, 2002	BFS	1652	0.10	0.08
z_4	edu	000s	Average number of schooling years per commune, 2004	BFS	1658	11.53	0.60

Fixed cost shifters

Variable	Name in dataset	Unit	Definition	Source	N	Mean (abs)	Std. dev. (abs)
W	farmlpc	CHF (100= 1/2000)	Area used for agriculture per capita for each commune, 1997	BFS	1667	0.86	2.19

Table 1 – Dataset used to estimate entry threshold ratios for Swiss newspaper sellers

The number of newspaper selling points in Swiss communes in the German-speaking part of Switzerland is analysed in this paper. This number is measured by the number of sellers which sell the national daily "Blick", which is one of the papers sold virtually in all distribution points in the German-speaking part of the country. It can be supposed that selling points which are selling this newspaper would also sell the relevant local newspapers of the regions in which they are active. It is therefore assumed to be a good approximation of the total number of newspaper sellers.

Figure 5 shows the frequency distribution for communes, by the number of sellers.

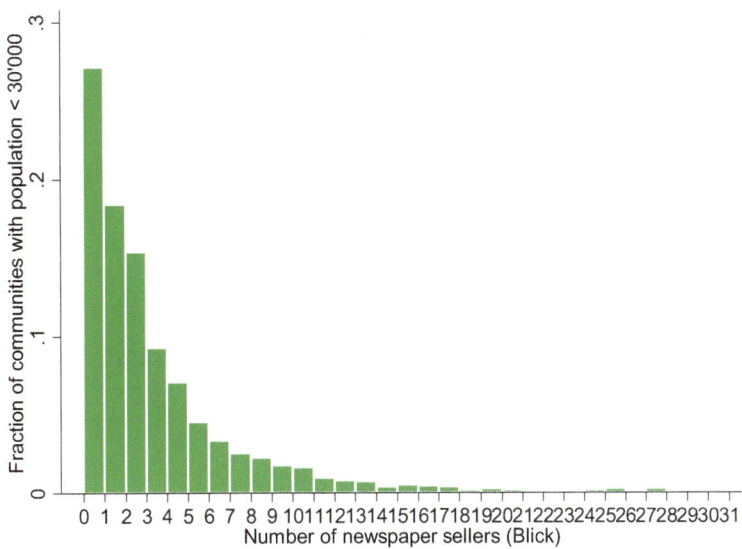

Figure 5 – Number of newspaper sellers in Swiss communes (frequency distribution)[34]

The number of observations corresponds to the total number of communes in the German-speaking part of Switzerland[35]. Many communes do not seem have a newspaper selling point (27%). All other communes have one or more selling points. Moreover, the case of more than four sellers in a commune is rare. There are, however, eleven outlier observations (Zurich for example is having 515 selling points). Also, as can be seen from Figure 6 the large majority of communities have a population below 1'000.

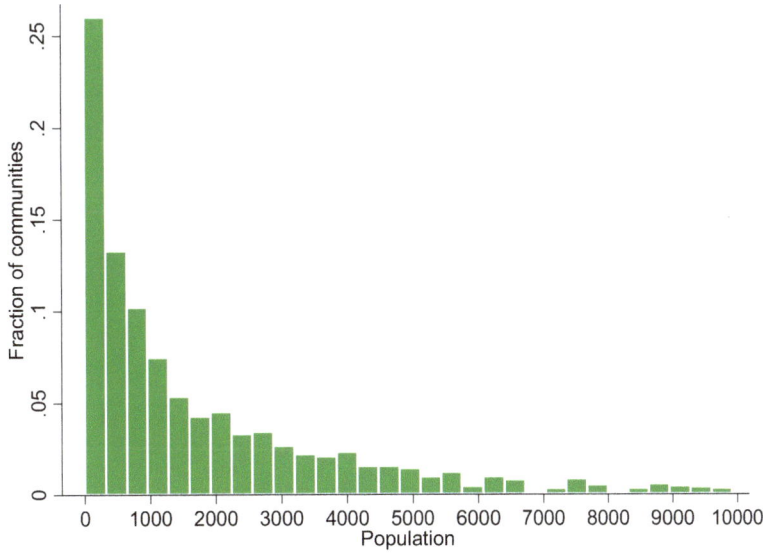

Figure 6 – Population in Swiss communes, frequency distribution[36]

---

[34] STATA input: "hist blick if pop<40000 & pop!=0"
[35] Except for the communes in the Canton of Thurgau, which were omitted, because there were changes in the geographic definition of communes over the years under consideration.
[36] STATA command: dotplot pop if pop<40000 & pop!=0

The model builds on the idea that population is one determinant of the relevant market size for newspaper sellers and that market size, in turn, influences the number of entrants. Residential population and inbound commuters to the commune are used as indicators for the size of the market a seller can potentially address in the commune[37]. Moreover, income is used as a demand shifting variable in the model. In addition, as in Bresnahan and Reiss (1991) demographic variables are used to control for possible differences between communes (population over 65 years, foreigners). Finally, the average schooling years are introduced as a control, as education levels may have an effect on demand for newspapers. Finally, it is proposed to include the farmland area per capita in the commune as an indicator for low real estate prices and, therefore as an adequate fixed cost shifter in a commune.

7. Estimation and interpretation of results

In this chapter , a model is fitted to calculate entry thresholds, per firm entry thresholds and per firm entry threshold ratios. Then, tests to verify the validity of results are performed and results are interpreted. Practically speaking, estimating Equation 19 is a complex task, which needs to be performed by designing a dedicated nonlinear ordered probit optimization algorithm (see annex) as well as test algorithms. Different similar algorithms have inspired this design, in particular Youle (2012) and Evdokimov (2013).

The regression results are reported in Table 2.

	Coefficient estimate	Std. err.	P>|z|
λ_0 (pop)	1 (set)	-	-
λ_1 (commuterIN)	.7810	.2173	0.000
β_1 (inc)	-.0028	.0014	0.050
β_2 (old)	1.494	.3138	0.000
β_3 (foreigners)	.5744	.1913	0.003
β_4 (edu)	-.0353	.0318	0.266
$V_1(\alpha_1)$	1.249	.3200	0.014
$V_1 - V_2(\alpha_2)$	.1652	.0673	0.020
$V_2 - V_3(\alpha_3)$	.1862	.0375	0.000
$V_3 - V_4(\alpha_4)$	.1103	.0245	0.000
$V_4 - V_5(\alpha_5)$	.1062	.0187	0.000
$F_1(\gamma_1)$	.4920	.0758	0.000
$F_2 - F_1(\gamma_2)$	.9882	.0818	0.000
$F_3 - F_2(\gamma_3)$	.3590	.0645	0.000
$F_4 - F_3(\gamma_4)$	.2120	.0597	0.000
$F_5 - F_4(\gamma_5)$	.0946	.0496	0.057
γ_R	-.0385	.0178	0.031
Prob> χ^2	0.000		
N	1596		

Table 2 – Nonlinear ordered probit estimates

---

[37] It is abstracted here from outward commuters.

As in Bresnahan and Reiss (1991), the coefficient on population was set equal to one (i.e. market size expressed in terms of residential population). Most coefficients are as expected and significant at 10%. Prob$(\chi^2) = 0.000$ indicates that the null hypothesis that all coefficients are jointly equal to zero can be rejected and therefore the model as a whole is significant as well. The following basic equations were implicitly estimated with the ordered probit model.

$$\hat{S}(Y,\lambda) = \widehat{\lambda_0} Y_0 + \widehat{\lambda_1} Y_1 \tag{20}$$

where Y_0 is population and Y_1 is inbound commuters.

$$\widehat{F_N} = \widehat{\gamma_1} + \widehat{\gamma_L} W + \sum_{n=2}^{N} \widehat{\gamma_n} \tag{21}$$

where the W variable is the available agricultural land per capita in the commune, and

$$\widehat{V_N} = \widehat{\alpha_1} + X\hat{\beta} - \sum_{n=2}^{N} \widehat{\alpha_n} \tag{22}$$

where the X variables include income in the commune and socio-demographic variables such as the number of aged people in the commune, the average number of schooling years and foreign resident population. As described before, it is expected that all α and γ's are positive in order to ensure decreasing average profits and increasing fixed cost with successive entry. This is the case for all 10 affected α and γ coefficients and they are all significant at 10%[38].

As described before, **CommuntersIN** is the number of people commuting to the commune under observation. Such commuters seem to have a significant and only slightly less strong importance in determining the relevant market size with respect to residential population (set to 1 for normalisation reasons). In this case, one resident of the commune counts as 1 in the market size considered by the newspaper seller. The commuters to the commune count as 0.78. Thus, the market size of a given commune can be calculated using the formula:

$$\hat{S}(Y,\lambda) = 1\,Pop + 0.78\,CommutersIN \tag{23}$$

This is important, as all subsequent analysis referring to the "market size" of a commune refer to this value for a given commune, not its population.

Income in the commune does not seem to be, as could be expected at first, a positive but a negative demand driver. This seems, however, reasonable as many goods typically sold by newspaper sellers are inferior goods (e.g. cold packaged sandwiches, tobacco, lottery, etc.). It is therefore possible that, in richer communes, people are less likely to visit such sellers and may prefer more specialised suppliers. In addition, it can also be expected that in such communes subscribership for newspapers is higher and demand at selling points is consequently lower.

The amount of **elderly people** in a commune seems to have a strong and significantly positive effect on the variable profits of newspaper sellers in a commune. This might reflect the fact that retired people have more time to visit such retailers. Also, the size of the **foreign** residential population seems to have a significant positive effect on newspaper sellers' profitability and entry. It is possible that this population segment (contrary to rich people) has higher demand for goods sold at typical newspaper selling points. Moreover, the number of average **schooling** years in a commune has an insignificant effect on

---

[38] The circumstance that γ_5 is not significant at 5% is unproblematic for the model. A late entrant might well once not have significantly higher fixed costs than the preceding entrant.

newspaper sellers' overall profitability. Finally, the **area** used for agriculture per capita is having the expected significant and opposite effect to land prices for which it is used as a proxy.

What is of particular interest in this article is, however, not the detailed demand or cost effects for products sold by firms selling newspapers, as these concern a largely undefined set. The focus of this article is on competitive effects of entry which are analysed in the next section.

Entry threshold estimation

The entry thresholds or minimum market sizes for an N-th firm to enter the Swiss newspaper sellers' market can be calculated using Equation 8, the ordered probit coefficient estimates and the mean values of the variables[39]. Results are reported in Table 3 and Figure 7. In theory, calculating the entry threshold might imply using state-dependent mean of the variables. It is, however, shown that in practice differences in results are negligible (these values are reported in the second rows of the table and are denoted by S^* (red in the figure) instead of S (blue in the figure)).

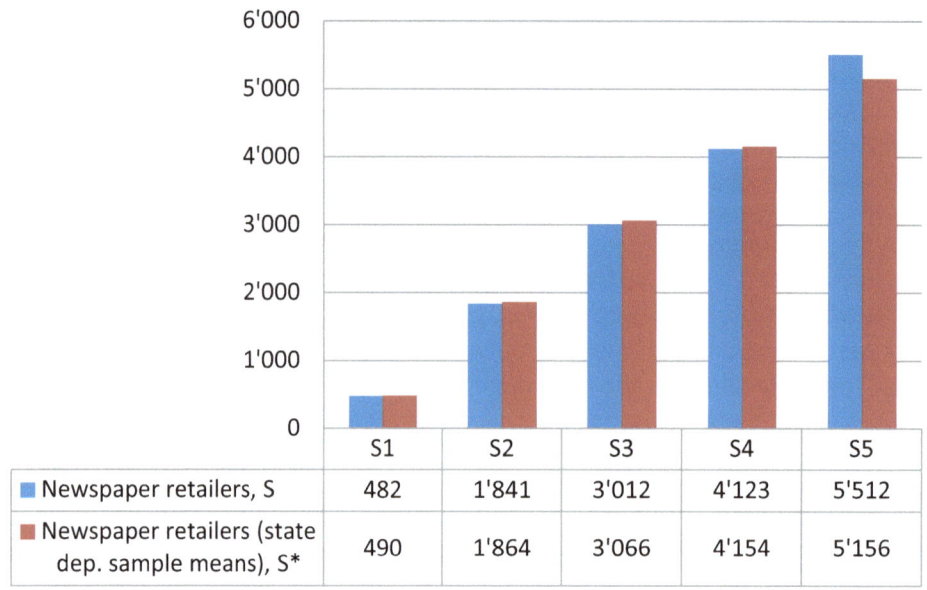

	S1	S2	S3	S4	S5
■ Newspaper retailers, S	482	1'841	3'012	4'123	5'512
■ Newspaper retailers (state dep. sample means), S*	490	1'864	3'066	4'154	5'156

Table 3, Figure 7 - Entry thresholds in terms of population[40], allowing an N-th firm to break even

Market size values are expressed here in terms of population (potential customers), but are referring to commune market size and therefore not only residential population. For example Table 3 indicates that a newspaper selling monopolist may enter at a market size of 482 (in terms of population), the second firm at market size 1'841 etc. Whether a given commune is below or above this threshold is calculated using Equation 23. For instance, when verifying whether entry of a first newspaper seller is economically sustainable in the town of Luchsingen, one would have to consider its population (938) as well as its inbound commuters (127). In total, this town would represent a market size of 1'037 (938 + 0.78*127). As the first entry threshold is 482 and the second is 1'841, it would be concluded that entry of one seller in this commune is economically viable, but that competition, i.e. the entrance of a second seller, is not. This conclusion is independent of the real situation in the single commune (whether entry has actually taken

---

[39] The STATA algorithm is programmed to calculate entry thresholds directly.
[40] While everything is expressed in terms of population, when checking whether a commune is affected or not by these rules, commuters need to be added to population.

place or not), which may also be different. It can be noted here that using state-dependent means[41] reduces the entry threshold for the 5th (or more) firm but does not otherwise lead to significantly different results.

Per firm entry thresholds

The per firm entry thresholds can be calculated using Equation 9 on the basis of the entry thresholds reported in Table 3. Per firm entry thresholds are reported in Table 4 and Figure 8.

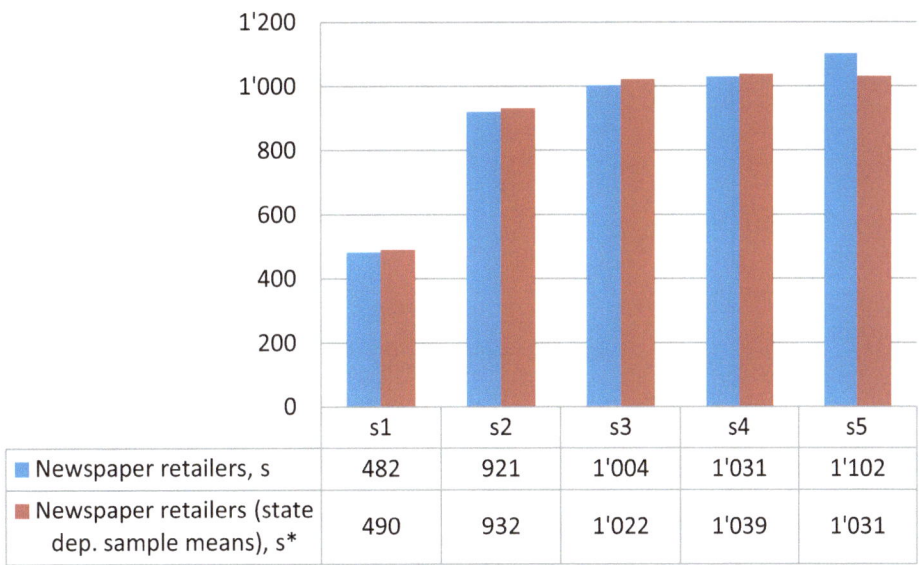

	s1	s2	s3	s4	s5
■ Newspaper retailers, s	482	921	1'004	1'031	1'102
■ Newspaper retailers (state dep. sample means), s*	490	932	1'022	1'039	1'031

Table 4, Figure 8 - Per firm entry thresholds in terms of population

It can be seen that per firm entry thresholds increase with entry as expected and settle around a per firm market size of 1'000 (in terms of population). It can be seen that the per firm entry threshold for a second firm to enter is nearly double the entry threshold for a monopolist. Competition therefore seems to strongly increase with the entry of a second firm. This result can be stated more clearly in the analysis of per firm entry threshold ratios.

Per firm entry threshold ratios

The per firm entry threshold ratios can be calculated using Equation 11 on the basis of the per firm entry thresholds reported in Table 4. Per firm entry threshold ratios are reported in Table 5 and Figure 9.

---

[41] This would imply, for example, that for the calculation of the estimates of variable profits (Equation 22) for each state (i.e. from zero to five entrants) the X values considered are not the overall sample means but the sample means applying to each state of the variable.

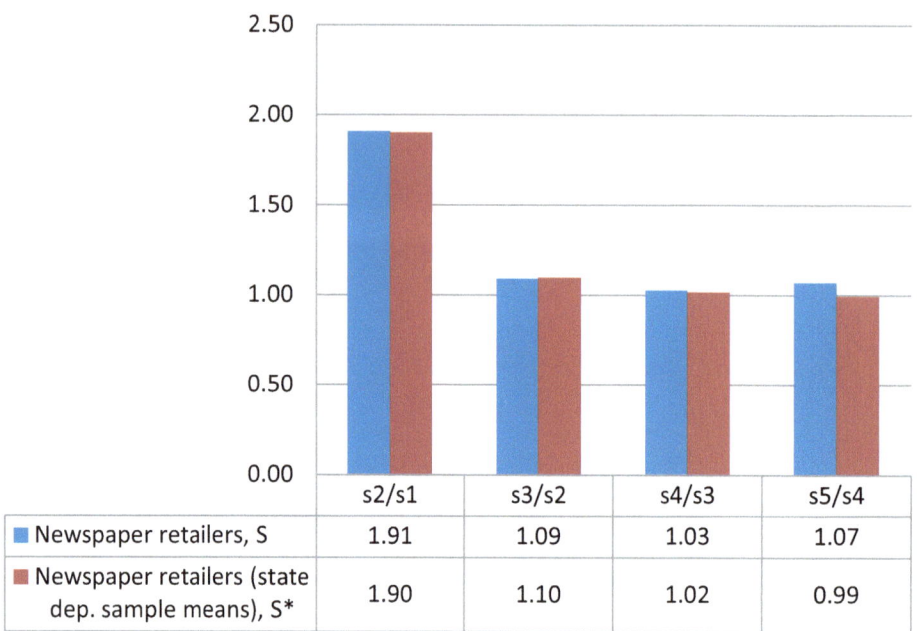

	s2/s1	s3/s2	s4/s3	s5/s4
■ Newspaper retailers, S	1.91	1.09	1.03	1.07
■ Newspaper retailers (state dep. sample means), S*	1.90	1.10	1.02	0.99

Table 5, Figure 9 - Per firm entry threshold ratios

The per firm entry threshold ratios behave as expected and approach one the more firms enter (meaning competition is not affected anymore by further entry). What must be noted here is that the first per firm threshold ratio s_2/s_1 is significantly higher than subsequent ratios. This seems to indicate that there is a strong increase in competition when the second seller enters. A slight increase in competition may be observed with the third seller, while entry of the fourth or further firms does not seem to further increase competition. It should be noted that when using full sample means s_5/s_4 is higher than the preceding threshold ratio. This should not be that case as this implies a slight reduction of competition with further entry. While this is theoretically possible it is unlikely in practice. The value of s_5/s_4 is, however, slightly different and in line with theory when using state-dependent means. It can therefore be expected that this effect is created by some outliers (larger cities) which are included in category 5. Using state-dependent means, such outliers may be addressed more appropriately.

Equality tests for per firm entry thresholds

In order to ensure that the conclusions on per firm entry threshold ratios and competition are correct a Wald test for threshold proportionality can be performed (Table 6). The algorithm is reported in the annex.

$s_5 = s_4$	$s_5 = s_4 = s_3$	$s_5 = s_4 = s_3 = s_2$	$s_5 = s_4 = s_3 = s_2 = s_1$
$\chi^2(1) = 0.08$ $P > \chi^2 = 0.77$	$\chi^2(2) = 0.21$ $P > \chi^2 = 0.90$	$\chi^2(3) = 1.42$ $P > \chi^2 = 0.70$	$\chi^2(4) = 35.32**$ $P > \chi^2 = 0.00$

Table 6 – Wald tests for equality of per firm entry thresholds

Formally, the test results show that the hypothesis $s_5 = s_4 = s_3 = s_2$ cannot be rejected. Conversely, it can be concluded that $s_5 = s_4 = s_3 = s_2 = s_1$ can be rejected.

The test results show that there is no significant difference between the per firm entry thresholds when moving from a duopoly to a three, a four or a five firm market environment. In other words, firm do not require substantially more customers to enter than in an environment with fewer competitors.

Unsurprisingly, the test confirms that the contrary is true when moving from a monopoly to a duopoly environment, where, as shown before, the market size necessary for entry of a duopolist to enter is much higher than the size necessary for the monopolist to enter.

The results imply equally that the hypothesis that each entry thresholds ratio pair is equal to 1 can be rejected except for the first pair. Again, this is not surprising as our graphical analysis shows that there is a strong decrease in per firm entry threshold ratios when the second firm enters (s_2/s_1), but not afterwards. When interpreting the right hand side of Equation 11 it can be concluded that when holding fixed costs constant, variable profits remain the same when moving from a two to a three, four or five (or more) firm environment, but not when moving from a one to a two firm environment. The test therefore indicates that there is a significant increase in competition with entry of the second firm, but none for subsequent entry. In other words, it seems that in the newspaper sellers' market two firms in a commune are sufficient to ensure competition.

Benchmarking results

It can be useful to compare these results to values found by other authors. While no other study has tried to estimate entry threshold ratios in the market for newspaper sellers, these results may be compared to results in other markets. Most importantly, the above results[42] can be compared to the values found by Bresnahan and Reiss (1991) for tire dealers, dentists, druggists and plumbers and to Abraham, Gaynor and Vogt (2007) for hospitals (Figure 10).

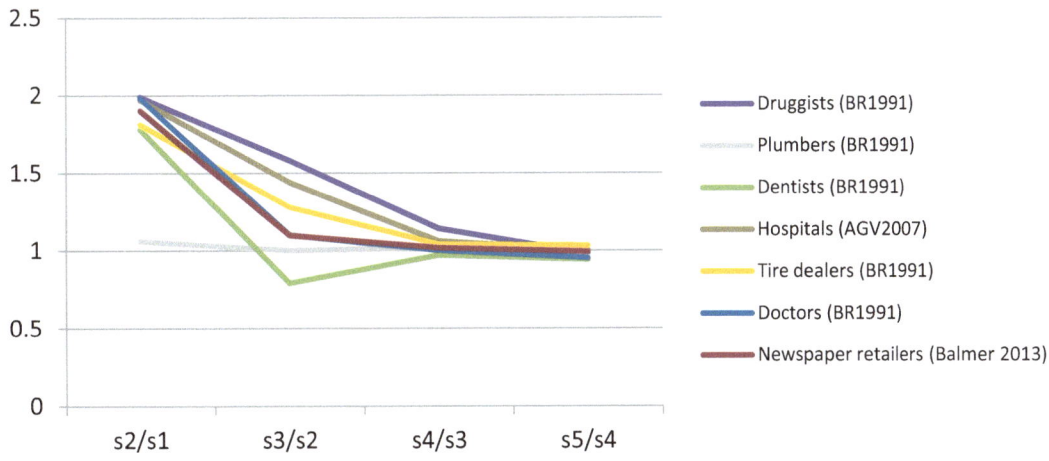

Figure 10 - Per firm entry threshold ratios for Swiss newspaper sellers in comparison

It can be seen that the estimates made in this article are well within the bounds of other estimates of per firm entry threshold ratios. Interestingly, Bresnahan and Reiss' (1991) estimates for U.S. doctors are virtually the same as those for Swiss newspapers. Entry effects on competition in these two markets can therefore be expected to be identical. Other results may also be stated here as the model has been applied by other authors in recent years. Most importantly, Abraham, Gaynor and Vogt (2007) study the entry of hospitals in U.S. cities. They find entry threshold ratios of 1.97, 1.44 and 1.06 meaning that competition is driven mostly by the entry of a second and third hospital (a development of competition with entry similar to druggists in Figure 10). Their results are used to conclude that blocking hospital mergers implying moving from three to two firms or from two to one in a local market (as contextually proposed by

---

[42] Using state-dependent means

the U.S. antitrust authorities but rejected by the Court) may likely cause significant harm to competition and consumers.

The estimate for the first entry threshold ratio in the Swiss newspaper sellers' market is high at 1.9 (i.e. the second entrant needs 90% higher per firm demand to enter than the first entrant). However, Abraham, Gaynor and Vogt (2007) and Bresnahan and Reiss' (1991) estimates of other markets - with the exception of plumbers - seem to be similarly high. It seems therefore that it is common that the second entrant importantly increases competition in the market. For subsequent entry, the situation is less clear and depends on the market. From a qualitative point of view, newspaper sellers can likely be best compared to (other) retail markets. For instance, a comparison with plumbers which usually sell services and may be more location independent may be less adequate. Also, entrance in the doctor or dentist markets can be expected to be highly regulated, which makes comparisons difficult. Results may, however, be well compared to tire dealers and druggists, both location based retailers. Most importantly, this comparison shows that while market power seems to decline relatively gradually with entry of the second and third firms for druggists[43] and tire dealers, this is not the case for Swiss newspaper sellers. The conclusion that two firms in a commune may be sufficient to ensure competition - or as the Dutch regulatory authority for telecommunications states "two are enough" (OPTA (2006)) - may therefore not apply in other retail markets.

Coverage

As has been shown before, the overall profitability of newspaper sellers strongly declines in the case of entry of a second newspaper seller in a commune. In what follows, the results in terms of coverage will be analysed.

If a monopolist in a small commune would earn only the variable profits a duopolist would earn, i.e. if it would behave as competitively as a duopolist, it would only enter in communes with more than 921 potential customers (duopoly entry level demand). This means that the additional margin extracted by a monopoly newspaper seller on the whole range of its products allows communes with a market size below 921, but above 482 (the entry level the monopolist needs to enter) to have a single selling point instead of having none. In Switzerland, concretely, according to the above, entry for newspaper sellers is possible for a first monopoly firm (with monopoly competitive conduct) only with a minimum of 482 potential customers. 310 (out of about 2'700) Swiss communes exist, where, for their market size, entry is in no case viable even for a monopoly applying monopoly prices. About 80'000 people are living in such (very small) communes, corresponding to about 1% of the Swiss residential population. In case the first entrant would only earn duopoly profits, it would only enter at a minimal market size of 921. In such case 263 additional communes, inhabited by 164'000 people or 2,1% of the population, would not be able to sustain a selling point. Thanks to the higher variable profits that monopolies can earn in the free market, a newspapers selling point in these communes is sustainable, and coverage can be extended by 2,1% of the population.

While less competition may increase prices and lower static welfare, our model shows that it may also lead to an important and clearly quantifiable effect on sustainable total coverage (dynamic welfare), which also needs to be taken into account in case of any public policy intervention (competition-coverage trade-off). From a welfare point of view, in a marginal area which would not be served under more competitive conduct by monopolists, monopolistic pricing would increase welfare. At the same time, such a pricing policy may decrease static welfare in other (monopoly) areas. Considering the overall effects, it is unclear which situation would be socially optimal (monopoly or duopoly prices in monopoly areas). Such a trade-off also arises, for instance, in the telecommunications industry, where a regulator has to set access

---

[43] It should be noted that in the U.S. druggists usually also sell newspapers.

prices in monopoly and duopoly areas to ensure not only local competition but also sufficient investment incentives for coverage (see Bourreau, Cambini and Hoernig (2012)). When considering only press products, where margins are independent of entry or even increasing with entry, a situation with monopoly conduct of newspaper sellers (over a whole range over products) in monopoly areas would likely be socially preferred. It should be noted, however, that the social benefit of coverage for press products may then come at the cost of, for instance, higher food or fuel prices at newspaper selling points. In terms of total coverage with press products only, it is beneficial that a monopolist earns relatively high margins. Similarly, though, it is optimal in terms of welfare that these margins decline strongly with a second entrant, as coverage is already ensured and there is little potential for differentiation in distribution in this model.

The model proposed also has some limitations. Firstly, consumers' travel costs are not addressed. Typically, the consideration of such costs would argue for increased industry demand with successive entry. At the same time the assumption of fully independent demand between communes may not be given as markets may overlap. Further empirical research is necessary to address these issues.

8. Policy Recommendations

Depending upon the priorities of legislators different public actions are appropriate in the single copy newspaper distribution market. Coverage of the country with selling points is not only important for citizens to inform themselves, but also for potential new newspaper entrants (editors) to reach (new) customers efficiently. There could therefore be reasons to think that the presence of the first seller in a local market has positive externalities for the economy and therefore should be supported.

Direct public intervention in the market

One way to ensure the availability of newspapers in local market is by direct public intervention. The government could instruct the Swiss Post to sell local newspapers (again) in some post offices, and, if necessary, to reduce other free market product sales. To maximize welfare access to the Post's sales infrastructure for publishers, wholesale distributors could also be granted access at regulated terms. As in telecommunications, though, a market intervention would likely only be justified in case of absence of (viable) competition. Otherwise, there is no need to enhance competition (and coverage) and any intervention would only distort competition. Absence of competition is at least given where in local markets entry of a newspaper seller is not viable. Our estimations show that this is the case in the 310 communes with a market size of less than 482. When looking at the actual data from the market, it can be seen that 405 communes inhabited by 210'000 or 2,7% of population are not served by a selling point in practice. Our model indicates, however, that in 95 more communes (covering about 130'000 inhabitants or 1,6% of population) entry would be economically sustainable.

The Swiss Post has intermittently sold newspapers in post offices. In 2007, it planned to offer newspapers in 470 post offices in the German-speaking part of Switzerland (out of about 2'600 in the whole country). Table 7 lists post offices where in 2007 selling points were set up (a full list was not published):

Trial post offices in 2007[44]	Estimation of S	No potential entrant	One potential entrant	Two potential entrants	Three potential entrants	Four potential entrants	Five or more potential entrants
		S<482	482<S< 1841	1841<S< 3011	3011<S< 4123	4123<S< 5512	S>5512
Aarau 1	32'519						X
Altdorf (UR) 1	11'918						X
Baden 1	31'650						X
Bern 1	203'368						X
Bern 14	203'368						X
Bern 18	203'368						X
Breitenbach (SO)	4'602					X	
Chur 4	41'540						X
Davos Platz 1	11'577						X
Domat/Ems	7'743						X
Grenchen 1	2'271			X			
Kreuzlingen 1	21'240						X
Landquart	8'719						X
Laufen	7'275						X
Luzern 1	87'718						X
Reinach 1	24'120						X
Romanshorn 1	11'896						X
Solothurn 2	27'711						X
Sulgen	4'358					X	
Winterthur 1	11'2819						X
Zug 1	39'017						X
Zürich 21 Sihlpost	510'052						X
Zürich 27 Enge	510'052						X
Zürich 47	510'052						X

Table 7 – Areas with post offices where selling newspapers; planned (Source: Swiss Post)

From the previous analysis, it is clear that, when in a commune no entrant is economically sustainable, public competition would only slightly distort the market by decreasing the potential profit of the following entrant, which is, however, unlikely to ever enter the market.

When analyzing entry in a market where one firm can be active, these considerations change. Firstly, there is a potential or even actual alternative firm. By entering this market, the Swiss Post would (following the assumptions in our analysis) get half of the market in the event that an actual competitor is present, and the full market when, even though potential entry is possible, in reality there is no firm present. In the first situation, the Swiss Post enters directly in competition with private firms on the free market. Such an intervention should only be conducted when there are potential benefits. In this case the only possible justification for such an intervention could be that a single seller cannot ensure sufficient competition, and that a public firm must enter to safeguard competition. Considering the results of this paper, this is likely to be only the case in a one firm market. In the second situation, the Post takes the place of a competitor which could otherwise have been a private firm. Finally, when two or more entrants are potentially possible, it can be seen that the competitive situation is vastly improved. In such case neither reason of coverage nor of competition could justify a market intervention by the government, as it is likely that the market is already competitive. From the list of trial post offices, it can be seen that all selling points were

---

[44] Source: Swiss Post

set up in communes where multiple selling points are economically viable. However, no plans for this business have been made in small communes, where typically no private firm could enter, but where post offices are available, or where only one firm is present. In light of earlier results such public policy seems therefore distortive of competition and is - under the assumptions made in this paper - unnecessary to increase coverage or competition.

Subsidies

In addition to direct market intervention public policy consisted and still consists in subsidizing small local editors by reducing the postal service prices they pay in order to reach their subscribers. If national coverage with newspaper sellers were seen as a priority, then such subsidization could be extended to early delivery for both subscriptions and for newspaper sellers' supply. Such reimbursement should occur independently of the delivery operator (today this includes only delivery by the Swiss Post to newspaper subscribers, and excludes early delivery).

Overall price levels

In January 2013, a survey by Swiss consumer organisations showed that international journals are sold at high prices at Swiss newspaper selling points. In particular, the study showed that journal and magazine prices were 58% higher when bought at a Swiss newspaper selling point rather than at a German newspaper selling point (e.g. magazines like "GEO" or "Spiegel")[45]. Comparison with France shows a similar difference of 59%. Among Italian newspapers popular in the Italian-speaking part of Switzerland, the difference was even 107%. The emphasis on this subject in public debate could put legislators and the competition authority under pressure to act in order to reduce final prices.

If institutions would want to increase price competition for newspapers at the retail level, price flexibility could be enforced. In areas in which there is more than one seller, competition between sellers on final prices might be enhanced. Conversely, under the assumptions of the model (newspaper sales are only a small proportion of total sales and margins), coverage of newspaper selling points would not be affected by this measure. In addition, the price difference with neighbouring countries might rather hint at competitive problems in the wholesale market (Valora). In the case of dominance, possible interventions include interventions based on competition law or direct market interventions in the wholesale market by the Swiss Post.

9. Conclusions

This paper estimates competitive effects of entry in the Swiss newspaper sellers market as well as coverage following Bresnahan and Reiss (1991). A market is considered where such sellers sell composite products including a broad range of other products such as food and near-food for which largely free price-setting on retail level is possible. The methodology adopted allows estimation without necessitating price and quantity data, therefore being useful for application in markets where data is rare and the range of products under examination not entirely clear.

It is found that a monopoly seller can enter communes with market size above 482 and two duopoly sellers can enter when the market size is above 1'841. All N-oligopolies would need each about 1'000 potential additional customers to enter. Entry is shown to be unviable in communes corresponding to 1% of the Swiss residential population, while entry in an additional 2.1% is shown to be only viable when

---

[45] preisbarometer.ch

monopolies are able to freely set monopoly prices on the range of products sold. There is therefore a clearly quantifiable trade-off between market prices and total coverage (investment).

The model additionally shows that competition increases strongly with the entry of a second newspaper seller in the market, while further firms cannot add significant further competitive pressure. It is therefore concluded that in the Swiss newspaper sellers' market, two firms are enough to ensure competition. Related articles have used such results to conclude that proposed mergers between firms, reducing firms in the market to not less than two, would not reduce competition or harm consumers. The results of this paper are, however, also compared to other retail markets and it is found that these conclusions may not be valid, for instance, for tire dealers or druggists, where also a third entrant may significantly enhance competition.

Finally, current public policy in this field is assessed and it is found, that the past local entry of government controlled Swiss Post in the newspaper sellers' market would have concerned only areas where the model predicts that viable competition between two or more sellers is sustainable. Such public policy is judged negatively, as it can neither enhance competition significantly nor extend coverage, but risks distorting competition. Nevertheless, it is shown that there are 310 communes where such a policy would not distort competition and could enhance welfare.

Bibliography

Abraham, J., Gaynor, M., & Vogt, W. B. (2007). Entry and competition in local hospital markets. *The Journal of Industrial Economics, 55*(2), 265-288.

Ahlers, D. (2006). News consumption and the new electronic media. *The Harvard International Journal of Press/Politics, 11*(1), 29-52.

Argentesi, E., & Filistrucchi, L. (2007). Estimating market power in a two-sided market: The case of newspapers. *Journal of Applied Econometrics, 22*(7), 1247-1266.

Argentesi, E., & Ivaldi, M. (2005). Market Definition in the Printed Media Industry: Theory and Practice: *C.E.P.R. Discussion Papers*.

Bignon, V., & Flandreau, M. (2011). The Economics of Badmouthing: Libel Law and the Underworld of the Financial Press in France before World War I. *The Journal of Economic History, 71*(03), 616-653.

Bignon, V., & Miscio, A. (2010). Media bias in financial newspapers: evidence from early twentieth-century France. *European Review of Economic History, 14*(3), 383-432.

Blasco, A., & Sobbrio, F. (2012). Competition and commercial media bias. *Telecommunications Policy, 36*(5), 434-447.

Bourreau, M., Cambini, C., & Hoernig, S. (2012). Geographic Access Rules and Investment. *CEPR-Centre for Economic Policy Research, Discussion Paper Series, 9013*, 1-48.

Bresnahan, T.F., & Reiss, P.C. (1991). Entry and competition in concentrated markets. *Journal of Political Economy, 99*, 977-1009.

Bundesamt für Kommunikation. (2012a). Häufige Fragen zur Presseförderung.

Bundesamt für Kommunikation. (2012b). Titel der Regional- und Lokalpresse.

Bush, C.A. (2002). On the Substitutability of Local Newspaper, Radio, and Television Advertising in Local Business Sales. *FCC Media Bureau Staff Research Paper, 10*.

Carlyle, T. (1993). *On heroes, hero-worship, & the heroic in history*: University of California Press.

Chamberlin, E. (1933). The Theory of Monopolistic Competition. *Cambridge, Mass, Harvard University*.

Dertouzos, J.N., & Trautman, W.B. (1990). Economic Effects of Media Concentration: Estimates from a Model of the Newspaper Firm. *Journal of Industrial Economics, 39*(1), 1-14.

Dewenter, R. (2003). The economics of media markets. *University FAF Economics Discussion Paper*(10).

Ecoplan. (2010). Evaluation der Presseförderung seit 2008 und alternativer Modelle.

European Commission. (2010). News Corp/BSkyB, COMP Case M.5932.

Evdokimov, P. (2013). Nonlinear ordered probit estimation algorithm for Stata. Retrieved from http://www.econ.umn.edu/~evdok003/mle.do

Filistrucchi, L., Geradin, D., Damme, E.v., & Affeldt, P. (2013). Market Definition in Two-Sided Markets: Theory and Practice. *TILEC Discussion Paper No.* (9).

Freedom House. (2012). *Freedom of the Press*.

Inderst, R., & Valletti, T. M. (2007). A tale of two constraints: Assessing market power in wholesale markets. *European Competition Law Review, 28*(2), 84.

Jefferson, T. (1787). *Letter to Colonel Edward Carrington*.

Latzer, M., Just, N., Metreveli, S., & Saurwein, F. (2012). Internet-Anwendungen und deren Nutzung in der Schweiz. Themenbericht aus dem World Internet Project: Universität Zürich.

M.I.S. Trend. (2009). Studie über den Internetanschluss und dessen Nutzung in der Schweiz.

Maddala, G.S. (1983). *Limited-dependent and qualitative variables in econometrics.* Cambridge University Press.

Montesquieu, C.d. (1751). The Spirit of the Laws. *11*, 69-70.

Nelson, P. (1970). Information and consumer behaviour. *The Journal of Political Economy, 78*(2), 311-329.

Newspaper Association of America. (2012). The Single-Copy Value Equation.

Office of Fair Trading. (2009). Newspaper and magazine distribution in the United Kingdom - Decision not to make a market investigation reference to the Competition Commission, OFT1121.

OPTA. (2006). Is two enough? Economic Policy Note 6. Retrieved from https://www.acm.nl/en/download/publication/?id=9102

Ordover, J. (2007). Comments on Evans & Schmalensee's 'The Industrial Organization of Markets with Two-Sided Platforms'. *Competition Policy International Journal, 3*.

Organisation for Economic Cooperation and Development. (2012). Market Definition. DAF/COMP(2012)19.

Panzar, J.C., & Rosse, J.N. (1987). Testing for monopoly equilibrium. *The Journal of Industrial Economics, 35*(4), 443-456.

Salop, S.C. (1979). Monopolistic competition with outside goods. *The Bell Journal of Economics, 10*(1), 141-156.

Shapiro, C., & Varian, H. R. (1998). *Information rules* (Vol. 29). Boston: Harvard Business School Press.

U.S. Department of Justice. (2011). Dynamic Competition in the Newspaper Industry, Remarks as Prepared for the Newspaper Association of America.

Valletti, T. (2006). Mobile Call Termination: a Tale of Two-Sided Markets. *Communications & Strategies*(61).

Van der Wurff, R. (2011). Are news media substitutes? Gratifications, contents, and uses. *Journal of Media Economics, 24*(3), 139-157.

Youle, T. (2012). Nonlinear ordered probit estimation algorithm for Stata. Retrieved from http://www.econ.umn.edu/~youle001/nloprobit.do

Annex: STATA12 algorithms to produce regression results, estimates and tests

STATA ".do" file to produce Table 2:

```
set more off
capture program drop nloprobit
program nloprobit
    args lnf S V1 a1 a2 a3 a4 a5 g1 g2 g3 g4 g5 F1
    tempvar P1 P2 P3 P4 P5
    qui gen double `P1'=normal(`S'*(`V1'+`a1')-`F1'-`g1')
    qui gen double `P2'=normal(`S'*(`V1'+`a1'-`a2')-`F1'-`g1'-`g2')
    qui gen double `P3'=normal(`S'*(`V1'+`a1'-`a2'-`a3')-`F1'-`g1'-`g2'-`g3')
    qui gen double `P4'=normal(`S'*(`V1'+`a1'-`a2'-`a3'-`a4')-`F1'-`g1'-`g2'-`g3'-`g4')
    qui gen double `P5'=normal(`S'*(`V1'+`a1'-`a2'-`a3'-`a4'-`a5')-`F1'-`g1'-`g2'-`g3'-`g4'-`g5')
    qui replace `lnf'=ln(1-`P1') if $ML_y1==0
    qui replace `lnf'=ln(`P1'-`P2') if $ML_y1==1
    qui replace `lnf'=ln(`P2'-`P3') if $ML_y1==2
    qui replace `lnf'=ln(`P3'-`P4') if $ML_y1==3
    qui replace `lnf'=ln(`P4'-`P5') if $ML_y1==4
    qui replace `lnf'=ln(`P5') if $ML_y1>=5
end

ml model lf nloprobit (lambda: commuterIN, nocons offset(pop)) ///
            (beta: inc old foreigners edu, nocons) ///
            /alpha1 /alpha2 /alpha3 /alpha4 /alpha5 ///
            /gamma1 /gamma2 /gamma3 /gamma4 /gamma5 ///
            (gammaL:blick=farmlpc, nocons)

ml search lambda 0 50 beta 0 1 alpha1 0 1 alpha2 0 1 alpha3 0 1 alpha4 0 1 alpha5 0 1 ///
            gamma1 0 1 gamma2 0 1 gamma3 0 1 gamma4 0 1 gamma5 0 1 gammaL 0 1
ml check
ml query
ml maximize, difficult

/*drop S1 S2 S3 S4 S5 s_1 s_2 s_3 s_4 s_5 s_21 s_32 s_43 s_54 betas ffracm lnhddm pincm eldm landvm*/
/*dist10kcity rent*/

qui egen incm=mean(inc)
qui egen oldm=mean(old)
qui egen edum=mean(edu)
qui egen foreignersm=mean(foreigners)
qui egen dist10kcitym=mean(dist10kcity)
qui egen rentm=mean(rent)
qui egen farmlpcm=mean(farmlpc)

qui gen betas=_b[beta:inc]*incm+_b[beta:old]*oldm+_b[beta:edu]*edum+_b[beta:foreigners]*foreignersm
qui                                                                                     gen
S5=[[gamma1]_cons+[gamma2]_cons+[gamma3]_cons+[gamma4]_cons+[gamma5]_cons+_b[gammaL:farmlpc]*farm
lpcm]/[[alpha1]_cons-[alpha2]_cons-[alpha3]_cons-[alpha4]_cons-[alpha5]_cons+betas]
qui                                                                                     gen
S4=[[gamma1]_cons+[gamma2]_cons+[gamma3]_cons+[gamma4]_cons+_b[gammaL:farmlpc]*farmlpcm]/[[alpha1]_c
ons-[alpha2]_cons-[alpha3]_cons-[alpha4]_cons+betas]
qui    gen    S3=[[gamma1]_cons+[gamma2]_cons+[gamma3]_cons+_b[gammaL:farmlpc]*farmlpcm]/[[alpha1]_cons-
[alpha2]_cons-[alpha3]_cons+betas]
qui gen S2=[[gamma1]_cons+[gamma2]_cons+_b[gammaL:farmlpc]*farmlpcm]/[[alpha1]_cons-[alpha2]_cons+betas]
qui gen S1=[[gamma1]_cons+_b[gammaL:farmlpc]*farmlpcm]/[[alpha1]_cons+betas]
```

```
qui gen s_5=S5/5
qui gen s_4=S4/4
qui gen s_3=S3/3
qui gen s_2=S2/2
qui gen s_1=S1/1

qui gen s_54=s_5/s_4
qui gen s_43=s_4/s_3
qui gen s_32=s_3/s_2
qui gen s_21=s_2/s_1

display " S1 " S1 _newline " S2 " S2 _newline " S3 " S3 _newline " S4 " S4 _newline " S5 " S5 _newline
_newline " s1 " s_1 _newline " s2 " s_2 _newline " s3 " s_3 _newline " s4 " s_4 _newline " s5 " s_5 _newline
_newline " s21 " s_21 _newline " s32 " s_32 _newline " s43 " s_43 _newline " s54 " s_54
```

Generation of estimates using state-dependent sample means (use after first .do file):

```
. qui gen betas1=_b[beta:inc]*incm1+_b[beta:old]*oldm1+_b[beta:edu]*edum1+_b[beta:foreigners]*foreignersm1
. qui gen betas2=_b[beta:inc]*incm2+_b[beta:old]*oldm2+_b[beta:edu]*edum2+_b[beta:foreigners]*foreignersm2
. qui gen betas3=_b[beta:inc]*incm3+_b[beta:old]*oldm3+_b[beta:edu]*edum3+_b[beta:foreigners]*foreignersm3
. qui gen betas4=_b[beta:inc]*incm4+_b[beta:old]*oldm4+_b[beta:edu]*edum4+_b[beta:foreigners]*foreignersm4
. qui gen betas5=_b[beta:inc]*incm5+_b[beta:old]*oldm5+_b[beta:edu]*edum5+_b[beta:foreigners]*foreignersm5
. qui gen
S5new=[[gamma1]_cons+[gamma2]_cons+[gamma3]_cons+[gamma4]_cons+[gamma5]_cons+_b[gammaL:farmlpc]*f
armlpcm5]/[[alpha1]_cons-alpha2]_cons-[alpha3]_cons-[alpha4]_cons-[alpha5]_cons+betas5
> ]
. qui gen
S4new=[[gamma1]_cons+[gamma2]_cons+[gamma3]_cons+[gamma4]_cons+_b[gammaL:farmlpc]*farmlpcm4]/[[alph
a1]_cons-[alpha2]_cons-[alpha3]_cons-[alpha4]_cons+betas4]
. qui gen
S3new=[[gamma1]_cons+[gamma2]_cons+[gamma3]_cons+_b[gammaL:farmlpc]*farmlpcm3]/[[alpha1]_cons-
[alpha2]_cons-[alpha3]_cons+betas3]
. qui gen S2new=[[gamma1]_cons+[gamma2]_cons+_b[gammaL:farmlpc]*farmlpcm2]/[[alpha1]_cons-
[alpha2]_cons+betas2]
. qui gen S1new=[[gamma1]_cons+_b[gammaL:farmlpc]*farmlpcm1]/[[alpha1]_cons+betas1]
. qui gen s_5new=S5new/5
. qui gen s_4new=S4new/4
. qui gen s_3new=S3new/3
. qui gen s_2new=S2new/2
. qui gen s_1new=S1new/1
. qui gen s_54new=s_5new/s_4new
. qui gen s_43new=s_4new/s_3new
. qui gen s_32new=s_3new/s_2new
. qui gen s_21new=s_2new/s_1new
. display " S1 " S1new _newline " S2 " S2new _newline " S3 " S3new _newline " S4 " S4new _newline " S5 "
S5new _newline _newline " s1 " s_1new _newline " s2 " s_2new _newline " s3
> " s_3new _newline " s4 " s_4new _newline " s5 " s_5new _newline _newline " s21 " s_21new _newline " s32 "
s_32new _newline " s43 " s_43new _newline " s54 " s_54new
```

STATA ".do" file to produce tests

```
. testnl ([ [gamma1]_cons+[gamma2]_cons+[gamma3]_cons+[gamma4]_cons+[gamma5]_cons+_b
[gammaL:farmlpc]*farmlpcm]/[5*[[alpha1]_cons-[alpha2]_cons-[alpha3]_cons-[alpha4]_cons-[alp
> ha5] _cons+betas]]=[[gamma1]_cons+[gamma2]_cons+[gamma3]_cons+[gamma4]_cons+_b[gammaL:farmlpc]
*farmlpcm]/[4*[[alpha1]_cons-[alpha2]_cons-[alpha3]_cons-[alpha4]_cons+betas]])
>        ([[gamma1]
_cons+[gamma2]_cons+[gamma3]_cons+[gamma4]_cons+_b[gammaL:farmlpc]*farmlpcm]/[4*[[alpha1]_cons-
[alpha2]_cons-[alpha3]_cons-[alpha4]_cons+betas]]=[[gamma1]_
> cons+[gamma2]_cons+[gamma3]_cons+_b [gammaL:farmlpc]*farmlpcm]/[3*[[alpha1]_cons-[alpha2]_cons-
[alpha3]_cons+betas]])  ([[gamma1] _cons+[gamma2]_cons+[gamma3]_cons+_b[gammaL
```

> :farmlpc]*farmlpcm]/[3*[[alpha1]_cons-[alpha2]_cons-
[alpha3]_cons+betas]]=[[gamma1]_cons+[gamma2]_cons+_b[gammaL:farmlpc]*farmlpcm]/[2*[[alpha1]_cons-
[alpha2]_cons+betas]])
>   ([[gamma1]_cons+[gamma2]_cons+_b[gammaL:farmlpc]*farmlpcm]/[2*[[alpha1]_cons-
[alpha2]_cons+betas]]=[[gamma1]_cons+_b[gammaL:farmlpc]*farmlpcm]/[1*[[alpha1]_cons+betas]]
>)

 (1) [[gamma1]_cons+[gamma2]_cons+[gamma3]_cons+[gamma4]_cons+[gamma5]_cons+_b
[gammaL:farmlpc]*farmlpcm]/[5*[[alpha1]_cons-[alpha2]_cons-[alpha3]_cons-[alpha4]_cons-[alpha5
>]_cons+betas]] = [[gamma1]_cons+[gamma2]_cons+[gamma3]_cons+[gamma4]_cons+_b[gammaL:farmlpc]
farmlpcm]/[4[[alpha1]_cons-[alpha2]_cons-[alpha3]_cons-[alpha4]_cons+betas]]
 (2) [[gamma1]
_cons+[gamma2]_cons+[gamma3]_cons+[gamma4]_cons+_b[gammaL:farmlpc]*farmlpcm]/[4*[[alpha1]_cons-
[alpha2]_cons-[alpha3]_cons-[alpha4]_cons+betas]] = [[gamma1]_c
> ons+[gamma2]_cons+[gamma3]_cons+_b[gammaL:farmlpc]*farmlpcm]/[3*[[alpha1]_cons-[alpha2]_cons-
[alpha3]_cons+betas]]
 (3) [[gamma1]_cons+[gamma2]_cons+[gamma3]_cons+_b[gammaL:farmlpc]*farmlpcm]/[3*[[alpha1]_cons-
[alpha2]_cons- [alpha3]_cons+betas]] = [[gamma1]_cons+[gamma2]_cons+_b[gammaL:
> farmlpc]*farmlpcm]/[2*[[alpha1]_cons-[alpha2]_cons+betas]]
 (4) [[gamma1]_cons+[gamma2]_cons+_b[gammaL:farmlpc]*farmlpcm]/[2*[[alpha1]_cons-[alpha2]_cons+betas]] =
[[gamma1]_cons+_b[gammaL:farmlpc]*farmlpcm]/[1*[[alpha1]_cons+betas
>]]

          chi2(4) =   35.32
        Prob > chi2 =   0.0000

. testnl ([[gamma1]_cons+[gamma2]_cons+[gamma3]_cons+[gamma4]_cons+[gamma5]_cons+_b
[gammaL:farmlpc]*farmlpcm]/[5*[[alpha1]_cons-[alpha2]_cons-[alpha3]_cons-[alpha4]_cons-[alp
> ha5]_cons+betas]]=[[gamma1]_cons+[gamma2]_cons+[gamma3]_cons+[gamma4]_cons+_b[gammaL:farmlpc]
farmlpcm]/[4[[alpha1]_cons-[alpha2]_cons-[alpha3]_cons-[alpha4]_cons+betas]])
>   ([[gamma1]
_cons+[gamma2]_cons+[gamma3]_cons+[gamma4]_cons+_b[gammaL:farmlpc]*farmlpcm]/[4*[[alpha1]_cons-
[alpha2]_cons-[alpha3]_cons-[alpha4]_cons+betas]]=[[gamma1]_
> cons+[gamma2]_cons+[gamma3]_cons+_b[gammaL:farmlpc]*farmlpcm]/[3*[[alpha1]_cons-[alpha2]_cons-
[alpha3]_cons+betas]]) ([[gamma1]_cons+[gamma2]_cons+[gamma3]_cons+_b[gammaL
> :farmlpc]*farmlpcm]/[3*[[alpha1]_cons-[alpha2]_cons-
[alpha3]_cons+betas]]=[[gamma1]_cons+[gamma2]_cons+_b[gammaL:farmlpc]*farmlpcm]/[2*[[alpha1]_cons-
[alpha2]_cons+betas]])

 (1) [[gamma1]_cons+[gamma2]_cons+[gamma3]_cons+[gamma4]_cons+[gamma5]_cons+_b
[gammaL:farmlpc]*farmlpcm]/[5*[[alpha1]_cons-[alpha2]_cons-[alpha3]_cons-[alpha4]_cons-[alpha5
>]_cons+betas]] = [[gamma1]_cons+[gamma2]_cons+[gamma3]_cons+[gamma4]_cons+_b[gammaL:farmlpc]
farmlpcm]/[4[[alpha1]_cons-[alpha2]_cons-[alpha3]_cons-[alpha4]_cons+betas]]
 (2) [[gamma1]
_cons+[gamma2]_cons+[gamma3]_cons+[gamma4]_cons+_b[gammaL:farmlpc]*farmlpcm]/[4*[[alpha1]_cons-
[alpha2]_cons-[alpha3]_cons-[alpha4]_cons+betas]] = [[gamma1]_c
> ons+[gamma2]_cons+[gamma3]_cons+_b[gammaL:farmlpc]*farmlpcm]/[3*[[alpha1]_cons-[alpha2]_cons-
[alpha3]_cons+betas]]
 (3) [[gamma1]_cons+[gamma2]_cons+[gamma3]_cons+_b[gammaL:farmlpc]*farmlpcm]/[3*[[alpha1]_cons-
[alpha2]_cons- [alpha3]_cons+betas]] = [[gamma1]_cons+[gamma2]_cons+_b[gammaL:
> farmlpc]*farmlpcm]/[2*[[alpha1]_cons-[alpha2]_cons+betas]]

          chi2(3) =   1.42
        Prob > chi2 =   0.7013

. testnl ([[gamma1]_cons+[gamma2]_cons+[gamma3]_cons+[gamma4]_cons+[gamma5]_cons+_b
[gammaL:farmlpc]*farmlpcm]/[5*[[alpha1]_cons-[alpha2]_cons-[alpha3]_cons-[alpha4]_cons-[alp
> ha5]_cons+betas]]=[[gamma1]_cons+[gamma2]_cons+[gamma3]_cons+[gamma4]_cons+_b[gammaL:farmlpc]
farmlpcm]/[4[[alpha1]_cons-[alpha2]_cons-[alpha3]_cons-[alpha4]_cons+betas]])
>   ([[gamma1]
_cons+[gamma2]_cons+[gamma3]_cons+[gamma4]_cons+_b[gammaL:farmlpc]*farmlpcm]/[4*[[alpha1]_cons-
[alpha2]_cons-[alpha3]_cons-[alpha4]_cons+betas]]=[[gamma1]_

> cons+[gamma2]_cons+[gamma3]_cons+_b [gammaL:farmlpc]*farmlpcm]/[3*[[alpha1]_cons-[alpha2]_cons-
[alpha3]_cons+betas]])

 (1) [[gamma1]_cons+[gamma2]_cons+[gamma3]_cons+[gamma4]_cons+[gamma5]_cons+_b
[gammaL:farmlpc]*farmlpcm]/[5*[[alpha1]_cons-[alpha2]_cons-[alpha3]_cons-[alpha4]_cons-[alpha5
>] _cons+betas]] = [[gamma1]_cons+[gamma2]_cons+[gamma3]_cons+[gamma4]_cons+_b[gammaL:farmlpc]
farmlpcm]/[4[[alpha1]_cons-[alpha2]_cons-[alpha3]_cons-[alpha4]_cons+betas]]
 (2) [[gamma1]
_cons+[gamma2]_cons+[gamma3]_cons+[gamma4]_cons+_b[gammaL:farmlpc]*farmlpcm]/[4*[[alpha1]_cons-
[alpha2]_cons-[alpha3]_cons-[alpha4]_cons+betas]] = [[gamma1]_c
> ons+[gamma2]_cons+[gamma3]_cons+_b [gammaL:farmlpc]*farmlpcm]/[3*[[alpha1]_cons-[alpha2]_cons-
[alpha3]_cons+betas]]

        chi2(2) =   0.21
        Prob > chi2 =   0.8984

. testnl ([[gamma1]_cons+[gamma2]_cons+[gamma3]_cons+[gamma4]_cons+[gamma5]_cons+_b
[gammaL:farmlpc]*farmlpcm]/[5*[[alpha1]_cons-[alpha2]_cons-[alpha3]_cons-[alpha4]_cons-[alp
> ha5] _cons+betas]]=[[gamma1]_cons+[gamma2]_cons+[gamma3]_cons+[gamma4]_cons+_b[gammaL:farmlpc]
farmlpcm]/[4[[alpha1]_cons-[alpha2]_cons-[alpha3]_cons-[alpha4]_cons+betas]])

 (1) [[gamma1]_cons+[gamma2]_cons+[gamma3]_cons+[gamma4]_cons+[gamma5]_cons+_b
[gammaL:farmlpc]*farmlpcm]/[5*[[alpha1]_cons-[alpha2]_cons-[alpha3]_cons-[alpha4]_cons-[alpha5
>] _cons+betas]] = [[gamma1]_cons+[gamma2]_cons+[gamma3]_cons+[gamma4]_cons+_b[gammaL:farmlpc]
farmlpcm]/[4[[alpha1]_cons-[alpha2]_cons-[alpha3]_cons-[alpha4]_cons+betas]]

        chi2(1) =   0.08
        Prob > chi2 =   0.7729

. ([[gamma1] _cons+[gamma2]_cons+[gamma3]_cons+_b[gammaL:farmlpc]*farmlpcm]/[3*[[alpha1]_cons-
[alpha2]_cons- [alpha3]_cons+betas]]=[[gamma1]_cons+[gamma2]_cons+_b[gammaL:farmlp
> c]*farmlpcm]/[2*[[alpha1] _cons-[alpha2]_cons+betas]])   ([[gamma1]
_cons+[gamma2]_cons+_b[gammaL:farmlpc]*farmlpcm]/[2*[[alpha1]_cons-[alpha2]_cons+betas]]=[[gamma1]_cons+_
> b[gammaL:farmlpc]*farmlpcm]/[1*[[alpha1] _cons+betas]])
unrecognized command: (invalid command name
r(199);

. testnl([[gamma1] _cons+[gamma2]_cons+[gamma3]_cons+_b[gammaL:farmlpc]*farmlpcm]/[3*[[alpha1]_cons-
[alpha2]_cons- [alpha3]_cons+betas]]=[[gamma1]_cons+[gamma2]_cons+_b[gammaL:
> farmlpc]*farmlpcm]/[2*[[alpha1] _cons-[alpha2]_cons+betas]])   ([[gamma1]
_cons+[gamma2]_cons+_b[gammaL:farmlpc]*farmlpcm]/[2*[[alpha1]_cons-[alpha2]_cons+betas]]=[[gamma1]_
> cons+_b[gammaL:farmlpc]*farmlpcm]/[1*[[alpha1] _cons+betas]])

 (1) [[gamma1] _cons+[gamma2]_cons+[gamma3]_cons+_b[gammaL:farmlpc]*farmlpcm]/[3*[[alpha1]_cons-
[alpha2]_cons- [alpha3]_cons+betas]] = [[gamma1]_cons+[gamma2]_cons+_b[gammaL:
> farmlpc]*farmlpcm]/[2*[[alpha1] _cons-[alpha2]_cons+betas]]
 (2) [[gamma1] _cons+[gamma2]_cons+_b[gammaL:farmlpc]*farmlpcm]/[2*[[alpha1]_cons-[alpha2]_cons+betas]] =
[[gamma1]_cons+_b[gammaL:farmlpc]*farmlpcm]/[1*[[alpha1] _cons+betas
>]]

        chi2(2) =   20.17
        Prob > chi2 =   0.0000

Chapter III

Competition and Market Strategies in the Swiss Fixed Telephony Market

An estimation of Swisscom's dynamic residual demand curve

Fixed telephony has long been a fundamentally important market for European telecommunications operators. The liberalisation and the introduction of regulation in the end of the 1990s, however, allowed new entrants to compete with incumbents at the retail level. A rapid price decline and a decline in revenues followed. Increased retail competition consequently led a number of national regulators to deregulate this market. In 2013, however, many European countries (including Switzerland) continued to have partially binding retail price regulation. More than a decade after liberalisation and the introduction of wholesale and retail price regulation, sufficient data is available to empirically measure the success of regulation and assess its continued necessity. This chapter develops a market model based on a generalised version of the traditional "dominant firm – competitive fringe" model allowing the incumbent also more competitive conduct than that of a dominant firm. A system of simultaneous equations is developed and direct estimation of the incumbent's residual demand function is performed by instrumenting the market price by incumbent-specific cost shifting variables as well as other variables. Unlike earlier papers that assess market power in this market, this paper also adjusts the market model to ensure a sufficient level of cointegration and avoid spurious regression results. This necessitates introducing intertemporal effects. While the incumbent's conduct cannot be directly estimated using this framework, the concrete estimates show that residual demand is inelastic (long run price elasticity of residual demand of -0.12). Such a level of elasticity is, however, only compatible with a profit maximising incumbent in the case of largely competitive conduct (conduct parameter below 0.12 and therefore close to zero). It is therefore found that the Swiss incumbent acted rather competitively in the fixed telephony retail market in the period under review (2004-2012) and that (partial) retail price caps in place can no longer be justified on the basis of a lack of competition.

1. Introduction

Fixed telephony was of fundamental importance to European telecommunications operators at the beginning of liberalisation at the end of the 1990s. In Switzerland, the incumbent Swisscom's fixed retail telephony revenues (access and traffic) made up for over 70% of its retail revenues in 1999[1]. In a context of overall declining telecoms revenues and fixed telephony prices, this share dropped to below 30% in 2011. Over this period, other telecommunications services such as broadband and mobile have increased in both relative and absolute importance. However, while a number of national regulators started to deregulate retail telephony markets after judging them competitive, in 2013, many European countries, including Switzerland, had partially binding retail price regulation for fixed calls still in place. More than a decade after the liberalisation and the introduction of wholesale and retail price regulation, sufficient data is available to empirically measure the success of regulation overall in recent years and to assess the continued necessity of retail price regulation.

This paper develops a market model based on a generalised version of the traditional "dominant firm – competitive fringe" model. Competitive fringe firms are assumed to be price takers and the dominant firm (Swisscom) acts as price leader able to move first, perfectly anticipating its competitors' supply reactions, to set an optimal unique market price for fixed telephony. The assumption of a "dominant firm" is subsequently relaxed considering also competitive conduct closer to price taking behaviour by introducing a *conduct parameter*.

A simple system of simultaneous equations is developed, allowing direct estimation of the incumbent's residual demand function by instrumenting the market price by incumbent-specific cost shifting variables, such as its number of staff (to control for possible remaining inefficiencies from times when the incumbent was fully state controlled), as well as other variables. Earlier papers have sometimes used indirect methods to estimate residual demand elasticity for a lack of good incumbent cost shifters (see Kahai, Kaserman and Mayo (1996)). The concrete model developed is, however, only valid under strong assumptions, including fixed termination rates close to zero, identical marginal network access costs for all competitors and independence from mobile telephony. Nevertheless, it is shown that these assumptions are reasonable in this context and the period under review (2004-2012).

Unlike earlier papers trying to assess market power in this market, this paper adjusts the market model to ensure a sufficient level of cointegration and to avoid spurious regression results. This necessitates introducing intertemporal effects. Any change in a variable has therefore an immediate same period effect on residual demand, and also a series of effects on future demand. While conduct cannot be directly estimated from this general framework, the concrete estimates show that residual demand is inelastic (long run price elasticity of -0.12). Such a level of elasticity is, however, only compatible with a profit maximising incumbent in case of largely competitive conduct (conduct parameter below 0.12 and therefore close to zero). It is therefore found that Swisscom acted rather competitively in the fixed telephony retail market in the period under review. If the problem of an uncompetitive retail market ever existed, it seems that the entry of alternative operators such as cable and the introduction of regulated wholesale access (carrier pre-selection as well as local loop unbundling) have successfully removed it. This implies that (partially)

---

[1] Swisscom's total fixed voice access and traffic revenues in 1999 amounted to 6.6bn CHF, compared to 1.8bn in mobile services and 0.9bn CHF in broadband. In 2011, fixed voice access and traffic revenues amounted to 1.8bn CHF, compared to 3.4bn CHF in mobile services and 0.6bn CHF in broadband (Swisscom annual reports 1999 and 2011).

regulated retail price caps in place in Switzerland can no longer be justified on the basis of a lack of competition, and calls into question whether there are legitimate reasons for their continuation.

Chapter 2 describes the fixed telephony market characteristics and the main assumptions of the model developed in this paper. In addition, it describes the concrete situation in Europe and in particular Switzerland, in terms of market structure and regulation. Chapter 3 outlines the basic analytical framework used to estimate fixed voice traffic (residual) demand. Chapter 4 provides an overview of the input data and Chapter 5 estimates the model. Chapter 6 interprets the results and Chapter 7 provides concluding remarks.

2. Market characteristics and model assumptions

This chapter describes the fundamental characteristics of telecommunications markets which need to be taken into account before building a market model for fixed telephony in Switzerland in subsequent chapters. In particular, the main properties of wholesale markets, where telephony operators buy access products from their competitors to originate and terminate calls, of cost functions and of demand are briefly analysed in light of the relevant literature. In addition, regulation and market structure over the relevant period is reviewed. From these characteristics, the main assumptions of the market model are derived.

Armstrong (2002) describes the fundamental properties of competition between retail fixed telephony operators (see also Laffont and Tirole (2001) and Vogelsang (2003)). Most importantly, operators are not identical. Competitors without infrastructure typically seek access to infrastructure owned by the incumbent to provide services to end-customers (one-way access). As the European incumbents usually are also present at the retail market, they are vertically integrated, while (most) competitors are only active on the retail market[2]. It is shown in this chapter that this situation is inefficient and can lead to foreclosure, which is why access is usually subject to some form of regulation. In addition, operators need to access each others' networks to terminate calls on competing networks (two-way access). While competitive interactions are more complicated, competition problems are also likely to arise, which is why termination is usually also subject to regulation. Wholesale regulation is therefore found to be one of the key drivers of the retail market for fixed telephony. Finally, it is shown that mobile calls do not represent good substitutes for fixed calls and that the fixed arms of firms with their own mobile networks take decisions independently in this context.

One-way access and vertical integration
Telecommunications operators need to have own physical infrastructure (local access network) or access to such an infrastructure to provide fixed telephone calls at users' homes. In most countries today, there are only few fixed access infrastructures allowing for independent fixed call origination (e.g. copper, cable, fibre) while at the time of liberalisation only one such infrastructure was broadly available (copper[3]). Operators without their own infrastructure may ask infrastructure-based operators for access[4]. In this case, so-called "one-way" access is requested, i.e. where the incumbent firm has control over an important input needed by

---

[2] The case of vertical separation is also analysed in Laffont and Tirole (2001). Typically the risk of exclusion does not exist (while the problem of monopoly power may persist).
[3] Cable networks were able to provide voice calls only later. In Switzerland, main cable operator Cablecom launched its telephone service in June 2004.
[4] It should be noted that wholesale access is usually not sold by cable operators.

its rivals to compete at retail level, but where it itself needs nothing from other firms. Competitors then pay an "access" charge to the incumbent. In other terms, Vogelsang (2003) relates access charges to cases where networks operate at different hierarchical levels and only one network uses the other to originate (and terminate) calls on the calling party's side. "Interconnection charges", instead, are related to two networks at the same hierarchical level, linked in order to enable calls across different networks (providing for example call termination for each other).

Returning to one-way access, the connection used to connect to the telephone network can technically be ensured by two different traditional wholesale access products (Armstrong (2002)). One is "call origination" where the incumbent provides for a telephone connection, and origination of traffic is bought by the competitor. In this case, the local telephone access continues to be operated by the incumbent and only long distance calls (i.e. over different regional points of interconnection where national calls are bundled) would require network elements operated by the access-seeking firms. A second possibility for entrants without their own access network is to provide telephony services to end-users through unbundling of the local loop, where the competitors operate their own telephone service platform up to the end-user[5].

Vogelsang (2003) reviews the relevant literature on one-way access and shows that under vertical integration, an incumbent may not be willing to grant one-way access, i.e. let competitors use the essential facility at reasonable terms, and that a foreclosure problem may exist[6]. Regulators should then impose access prices. Optimal access prices maximising welfare under the constraint that the incumbent breaks even are called Ramsey prices, and are typically above marginal costs as in this case the competitors have to make a fixed cost contribution to the incumbent. Given the complexity of implementing Ramsey pricing (e.g. lack of information on elasticities and competitive reactions), such pricing has not yet broadly been implemented by regulators. Instead, telecom regulators have usually used cost-based access price determination mechanisms (in particular long run incremental cost (LRIC) where the investment costs for the different services are shared with rivals on cost basis). The level and impact on the market of a regulated fixed call origination charge is analysed in the next section.

Call origination costs

The telephony operators' total as well as marginal costs can be divided in two broad categories. One is the network costs faced by a firm to establish the necessary telephone connections and handle the traffic and the other is costs such as customer care and administration. For alternative operators without their own network, the network access costs are represented mainly by the (potentially above marginal cost) regulated access prices. The incumbent a priori faces only its marginal cost curve. With price independent demand this may be irrelevant for the market outcome as the incumbents' perceived marginal cost corresponds to its opportunity cost on the wholesale market. With price dependent demand[7] (and when each competitor has a price dependent hinterland of loyal customers unaccessible to the other operator), instead, this can lead to inefficiencies. In this case, when an incumbent is vertically integrated and retail competitors rely on its wholesale products, Inderst and Peitz (2012) show that the incumbent charges lower uniform retail prices in equilibrium than its competitors and that it has a higher market share (partial

---

[5] In Switzerland markets referring to both mentioned access products were found to be not effectively competitive, respectively the incumbent has never denied to have significant market power, and regulation is actively imposed (LRIC prices).
[6] With differentiated goods this problem may partially persist when demand is price dependent (see Inderst and Peitz (2012)). In fact, the incumbent would set an access price above marginal cost and would set lower retail prices than the competitor in order to attract more customers in the loyal customer segment where it faces no opportunity cost but only its marginal network cost.
[7] Which can be assumed to be the case

foreclosure)[8]. This would be the case whenever regulated access prices are above marginal network cost[9]. Conversely, Gans and King (2005) show that when access prices for non integrated competitors are set at marginal network cost (and integration does not provide strategic benefits or costs) this implies competitive neutrality. When the regulated access price is set at marginal network costs, the competitor can therefore compete at exactly the same grounds as the incumbent and there is a priori no reason for inefficient allocation even in case of product differentiation and price dependent demand. When there are two firms, the retail market outcome after investment is then equivalent to a competitive model of a market without vertical integration.

The relevant one-way access prices in Switzerland are regulated at cost-based LRIC prices (i.e. above marginal network cost of the incumbent). Access prices are required to be non-discriminatory, meaning that the incumbent's retail arm faces a marginal network cost equal to LRIC prices - as all its access seeking competitors. A drop in the cost and price of the regulated product would therefore have the effect to reduce perceived marginal costs in an identical way for both the incumbent and the competitors. The costs taken into account by the cost model of the regulator for "regional origination" of calls are the elements to construct a telephone network and originate calls (not considering the construction of the physical access network which is a separate increment). Essentially, these costs are related to the operation of a local and regional telephone switching platform and corresponding backhaul lines. It can be assumed that, differently to the physical access network, a significant part of these costs are operating and maintenance costs and that such costs may depend on scale. In this particular case, the regulated long run incremental costs are therefore assumed to be close to marginal costs (of an efficient operator). In the model it is therefore assumed for simplicity that LRIC prices for this wholesale product correspond largely to marginal network cost of origination and that access seekers therefore compete on the same grounds as the access providers when considering call origination. The aspect of vertical integration of the incumbent can therefore be ignored.

Competitors with their own infrastructure have not been considered in this analysis. In Switzerland, cable operators started to offer their own telephony services in 2004. In addition, competitors started to provide telephony services based on local loop unbundling in 2008. However, it can reasonably be assumed that all these operators are similarly efficient in operating telephone networks as an efficient incumbent[10], and that they therefore face the same marginal costs for call origination and have the same marginal network cost drivers. Such costs are in this paper assumed to be largely exogenous and to be given by the regulated access price for call origination. Additional network costs to complete calls (conveyance and termination on different networks) as well as other costs related to the operation of a telephony networks (core network maintenance, customer care, administration, etc.), however, also need to be considered by competitors.

Marginal network costs are therefore considered identical for origination for all operators and exogenous. All other costs are supposed to be able to differ across operators (e.g. long distance network elements, customer care, etc.). For overall marginal costs, a positive slope is assumed in the segment under consideration. In other words, as in most markets, it is assumed that at some point diseconomies of scale set in and marginal costs increase, essentially due to limited capacity. When some threshold of traffic is reached, operators may have to install more backhaul capacity which may be very costly and lead to very

---

[8] As the incumbent in its hinterland incurs only its marginal network cost and not an (above marginal cost) opportunity costs corresponding to the other operators' wholesale cost, an asymmetry between the firms arises.
[9] It should be noted that there can also be a so-called "softening effect". When upstream access is provided a customer lost on the retail market may then be recovered partially via the upstream market. This may make integrated firms less aggressive on the retail market than other firms.
[10] The regulators costs are calculated for an efficient incumbent, therefore considering always the most efficient technology to provide the service.

high marginal costs for the level of demand which could trigger such an investment. Similar considerations can be made for other costs. For example, the main Swiss fixed telephony competitor Sunrise seemed to face an important increase of costs at some point. The increase in demand has made it necessary for the firm to recruit about 170 new call center operators (on 830 existing[11]) in only three months and to buy additional (detached) office spaces. For these reasons it is therefore plausible that for the operation of a telephone network alone, overall marginal costs of the operators may be increasing.

It can therefore be concluded that the assumptions that the hypotheses' that:

i) regulated LRIC origination prices correspond to an efficient operator's marginal network cost (and all infrastructure-based operators are assumed to be efficient)
ii) overall marginal cost is increasing with output

are plausible in the context under analysis implying that operators compete on the same grounds in the retail market and that competitive firms would increase their output with the market price.

Two-way access

In order to provide a successful phone call from a users' home, an operator not only needs access to the customer, but must also ensure conveyance of its calls and delivery by the competitor providing services to the user called (termination). Under two-way access, of which termination is an example, operators use each others' bottleneck inputs to be able successfully terminate calls and compete with each other. The need for regulation is here less straightforward than under one-way access. Armstrong (2002), Briglauer (2010), Valletti and Calzada (2005) and Valletti and Genakos (2011) propose market models taking into account termination costs and revenues. The literature review of Vogelsang (2003) shows that two issues may arise: collusion (especially under symmetric interconnection relationships) and exclusion (especially when relationships are strongly asymmetric). While every operator may have exclusive control of calls terminated on its subscriber base it must at the same time reach an agreement with competitors on a termination rate in order for its customers to be able to call the competitors' subscriber base. The equilibrium outcome largely depends on the operators' positions and market parameters. As both collusion as well as exclusion undermine competition, this situation has led in Europe to price regulation of termination rates (usually cost-based).

In Switzerland, termination on the incumbent's fixed network is similarly actively price regulated (LRIC) since (nearly) the beginning of liberalisation. While there is no direct regulation of the other fixed operators' termination fees, the incumbents' regulated fixed termination standard offer for fixed operators contains a reciprocity clause[12]. Such clause means that bilateral termination rates of Swisscom with its fixed competitors are identical and - as the incumbent also offers regulated transit to fixed termination of third operators[13] - that all other fixed termination rates are indirectly regulated by arbitrage at or near the incumbents' rate. It is therefore assumed that all operators receive the same regulated fixed termination fee. It should be noted here that from a practical point of view only operators with independent telephone networks receive termination fees. Therefore, for telephone operators without an own network as via unbundling or cable, the termination fees are received by the network operating firm (the incumbent).

---

[11] Source: http://www.20min.ch/digital/news/story/Bei-Sunrise-sitzt-der-Chef-im-Call-Center-27828675
[12] Source: http://www.swisscom.ch/en/wholesale/products/voice-services.html and
http://www.swisscom.ch/dam/swisscom/de/ws/documents/D_IC-Dokumente/D_IC_LB_Terminierungsdienste_V1-0.pdf
[13] Swisscom Transit Termination

No active regulation is, however, in place for mobile termination in Switzerland. This is due to the litigation based regulatory system, where a market intervention is only be triggered upon request by an operator. There has never been an upheld lawsuit of an operator against another operator to ask for regulated access to termination on its mobile network yet. Therefore such termination rates are still unregulated. Unsurprisingly, Swiss mobile termination rates were (among) the highest in Europe in the period under consideration. For example in January 2010, Switzerland had the highest average mobile termination rate, 10.7€cent per minute, compared to a European average of 6.3€cent[14]. For comparison, the price of the regulated fixed termination charge in Switzerland was then significantly below 1€cent per minute[15]. Mobile termination charges were in absolute terms therefore more than ten times higher than fixed termination charges. This particular regulatory framework may imply that the cost to terminate mobile calls was of far greater importance to fixed operators' profits than any possible (fixed) termination revenue[16], especially when considering a homogenous good of voice traffic towards all national networks. Therefore, when considering an overall market for national calls, a strongly simplified framework - necessary as will be shown for estimation reasons - could for the above reasons foresee full abstraction from fixed termination revenues and costs of fixed operators, while taking into account mobile termination rates on the cost side.

It can therefore be concluded that the assumptions that:

i)   all Swiss fixed operators face the same fixed regulated termination charge
ii) the fixed termination charge is zero and can be ignored in the model in terms of revenues and costs
iii) the mobile termination charges are high and taken into account as costs for fixed operators

are plausible, implying that all fixed operators compete on the same grounds in the retail market.

Horizontal differentiation

Unlike other telecommunications services, telephone calls from the fixed network usually have no particular features that allow for horizontal differentiation, and vertical differentiation seems largely excluded as well[17]. The differentiation characteristics considered in Kahai, Kaserman and Mayo (1996) referred to the so-called "carrier selection" model, where a customer wishing to use an alternative operator had to dial additional digits before the telephone number (this could be interpreted as negative quality). However, with the introduction of "equal access" wholesale offers in the U.S. ("carrier pre-selection"), no additional digits were necessary anymore for the preselected carrier. By 1993, in the U.S. 97% of wholesale offers were converted to equal access. In Switzerland in the period under review, carrier preselection could be opted for by competitors, foreseeing that customers would not have to dial additional digits. Such source of differentiation is therefore absent in this case and it can be assumed that fixed telephony is a largely homogeneous good. In the case of a homogeneous good, a single equal retail price for telephone services can be assumed as otherwise rational consumers would tend to switch. This assumption may be limited by the fact that switching may be complicated and lengthy or by differentiation of brands. Only in recent years - and therefore unlikely to affect the present analysis of the years 2004-2012 - a further source of differentiation may be present with bundles (in particular where some operators can offer additional services

---

[14] Source: BEREC, BoR(10)30rev1
[15] Source: Swisscom Price Manual 7-2
[16] This can also be seen, for example from the case of Sweden, where the regulatory authority has published industry fixed voice service revenues in the market (including fixed charges, fixed to fixed and fixed to mobile calls, international calls and other) as well as fixed termination revenues. Fixed termination revenues corresponded to less than 5% of the service revenues in 2012. Source: PTS: "The Swedish telecommunications market 2012"
[17] Commercialised HD calls and video calls are still rare.

such as TV and others not). In addition, on the market there are a number of price plans and components and comparability and transparency may not be a given (e.g. two-part tariffs, on- and offnet tariffs, day and night/weekend tariffs, bucket subscriptions, etc.). Nevertheless, in light of the above it seems reasonable to assume for simplicity that fixed telephony services are a homogeneous good. This paper abstracts therefore from residual differentiation possibilities.

Fixed-mobile substitution

When designing a market model to estimate market power in the fixed telephony market possible substitutes need to be taken into account. Fixed telephony has one major potential competitor, which is mobile telephony. Typically, mobile access and calls are more expensive than fixed network calls, as they provide the feature of mobility. Vogelsang (2010) reviews the literature on fixed-mobile substitution and concludes that mobile telephony is a substitute for fixed telephony (positive cross-price elasticity) in wealthy countries[18] (calls only, while the situation is less clear for access). Moreover, the author argues that with increasing mobile penetration and decreasing prices, substitutability should further increase over time. While theoretical work is often inconclusive, empirical work, most prominently Briglauer, Schwarz and Zulehner (2010) hints towards substitution. The authors analyse monthly telephony market data in Austria from 2002 to 2007. They find that for residential users, there is a positive cross-price elasticity between fixed and mobile for national calls. They consequently argue therefore that fixed and mobile calls should be considered part of the same market. However, the Austrian market is notoriously very competitive and results do not have to translate to Switzerland under the period of review. Most importantly, the Austrian regulator was to date the only European regulator finding a joint retail market for fixed and mobile voice telephony (Austrian regulatory authority case AT-2009-0881). All other European regulators have to date not come to this conclusion in their national markets. A survey by BEREC (2012) shows that the main reason for these regulators not to define a common retail market was the existence of different product characteristics between fixed and mobile offers, in particular different price levels and the mobility of mobile services. BEREC (2012) cite an Analysys Mason study[19] estimating that in Western Europe fixed calls are cheaper by 37% than equivalent mobile calls. In addition, the conclusion of Briglauer, Schwarz and Zulehner (2010) does not apply to access. In this case, it is argued that probably for quality differences and the possibility to share costs among household members there seems to be no fixed mobile-substitution.

Given the above, a model of the fixed telephony market (calls) should probably also take into account supply and demand for mobile calls. When considering a long period of time retrospectively and in countries with particularly high mobile prices or low mobile penetration and, this may, however, not be necessary.

Regarding prices, a study by the Finnish Communications Regulatory Authority (2009) finds (at about mid-period of the dataset which will be considered) that from 19 European countries, Switzerland had in by far the highest mobile telephony prices with a medium basket expenditure of around 70€ per month compared to a European average of 42€ per month. Conversely, fixed telephony prices were closer to the average[20]. As in addition a long retrospective period is considered (2004-2012), the present model can reasonably assume the absence of fixed-mobile substitution. A factor supporting this hypothesis is that the duration of

---

[18] Where the fixed network coverage is not more extended.
[19] Mobile-only households: fixed voice will all but disappear in some Central and Eastern European countries – September 2010
[20] OECD Communications outlook 2011. E.g. 140 fixed calls, VAT included, Switzerland: 26$, OECD average: 27$

fixed calls in Switzerland was found to be more than double the duration of mobile calls before 2010, indicating a different use of the two technologies[21].

Independence of fixed telephony

Even in a framework where absence of fixed-mobile substitution is assumed, it is, however, not clear whether the fixed and mobile markets are fully independent. Mobile operators have revenues from mobile termination fees paid by fixed operators. When an integrated operator offers mobile and fixed network services it might, when setting a fixed retail price, internalise effects of mobile termination revenues. For instance, lower fixed voice prices and higher volumes in the market could increase an operators mobile termination revenues. In practice, in the case of Swisscom, its high mobile termination rate had implied that its fixed division payed consistent termination fees to its mobile division[22]. As both fixed operators with mobile networks had separate business units for fixed and mobile networks with separate objectives for a large part of the period under analysis, it is assumed in this paper for simplicity that fixed operators operate independently even when they have mobile arms. They are therefore assumed to pay mobile termination rates as any other operator and to set prices to maximise their business unit profit. In addition, it can be assumed in this case that fixed operators perceive mobile termination rates as (largely) exogenously given, as for example in Briglauer, Götz and Schwarz (2010).

Retail price structure

In this section it may be convenient to review market prices before reviewing the theory on different tariff structures. As in most countries, standalone telephony services for residential customers in Switzerland includes telephony access and a corresponding access fee, as well as usually standard call set-up and per minute charges. Telephone access has in older studies been shown to be price inelastic. For a review of the literature on fixed voice access own price elasticities see Ward and Woroch (2010) and Gassner (1998). It can therefore be assumed that competition works mainly via the usage-based part of the tariff. This assumption seems to be confirmed in the Swiss market where in 2013 a standalone telephone access costs about 25 CHF per month at all major fixed telephone operators (Swisscom, Sunrise, Cablecom). This retail access fee corresponds to the maximum charge for universal service (including VAT) as defined in the Swiss Telecommunications Act[23] and in particular the Ordinance on Telecommunications Services[24] (article 22). Volume based tariffs differ, however. For example, Swisscom charges 0.08 CHF/min for calls to the fixed network (50% of it during the night and on weekends) and 0.35 CHF/min for calls to the mobile network (0.30 CHF/min during the night and on weekends). Swisscom's per minute charge again corresponds to the maximum charge for universal service. Cablecom, instead, foresees no charges for calls to the fixed network while it charges 0.40 CHF to all mobile networks. Sunrise charges 0.06 CHF/min to the fixed network (free during the night and on weekends) and 0.35 CHF to mobile networks (0.30 during the night and on weekends). In addition, these calls incur call setup fees. In particular, Cablecom and Sunrise charge a setup fee of 0.12 CHF per call, while Swisscom does not charge such a fee. In summary, all major operators charge the same fixed fee, but differentiate their price plans according to call set up fee, day and night/weekend tariffs, fixed and mobile calling prices, and a free initial number of minutes. In addition, a number of subscriptions are available.

---

[21] For example, in its 2010 decision on mobile termination the Swiss federal court of Justice (B-2050/2007) argued that the fixed call duration in the period under review was more than double the mobile call duration. This seemed to clearly indicate a different use of the two technologies, the more expensive, mobile solution on the go and the cheaper stationary solution at fixed locations.
[22] The Swiss federal court states that in 2005 Swisscom (Fixed division) had paid transfer payments to the three mobile operators (including Swisscom mobile) of more than 100m CHF. See RPW 2010/2.
[23] http://www.admin.ch/opc/en/classified-compilation/19970160/201007010000/784.10.pdf
[24] http://www.admin.ch/opc/en/classified-compilation/20063267/201212280000/784.101.1.pdf

The simplest possible model setup regardind the retail tariff structure would be a linear tariff. This would mean the absence of a fixed fee (two-part tariff) and the presence of a single usage-based charge per minute (i.e. taking into account any call setup charges, day and night/weekend charges, all fixed and mobile calling prices and free minutes). As can be seen in the next chapter the number of telephone accesses (even excluding Internet based telephone services such as Skype) is by far exceeding the number of households in Switzerland (about 5m accesses on average during the period under review against about 3.2m households). It can therefore be assumed that - even though there may be a number of business lines - in a large majority of households an active telephone access was present. In addition, demand for access is typically inelastic. It is therefore reasonable to think that competition for traffic largely independent from access.

Nevertheless, a two-part tariff nature of pricing in fixed telephony markets may have indirect effects. The presence of a fixed fee may correspond, in fact, to some level of transfer (based on bargaining power) between consumers and firms, while the traffic based prices aim to maximise rent extraction (see for example Inderst and Peitz (2012)). Of relevance for this particular case, Growitsch, Marcus and Wernick (2010), have shown that the fixed part of the tariff has a strong negative correlation with the level of the (fixed) termination rate. To see this, the literature on "waterbed" effects has to be considered. Termination brings not only costs for operators but also revenues. The revenues a consumer brings to an operator from other users calling him and terminating on its operators' network may be relevant. When such revenues are important competition may imply that an operator (e.g. a mobile operator with high termination charges) may lower its retail prices to attract consumers in order to have access to the related termination revenues. Some authors argue that this "waterbed" effect mostly affects the fixed fee, which is reduced in such cases. Conversely, when termination rates are lowered by regulatory authorities, termination revenues per user are more limited and the possibility of granting discounts to attract customers on this basis are reduced. Overall, it is unclear whether cost or revenue effects will dominate. Some literature has developed to understand the extent of the waterbed effect. For instance, Genakos and Valletti (2011) estimate that lower mobile termination rates would lead to higher overall mobile retail prices. Most authors find, however, that lower termination rates lead to a decline in retail prices but to a lower extent (implying a partial or incomplete waterbed effect). This does not mean that all components of the retail price would decrease. Most authors, like Growitsch, Marcus and Wernick (2010), assume that the fixed fee would be increased in case of a reduction of termination rates (negative correlation). However, as in this model also the absence of (relevant) fixed termination charges and fixed termination revenues is assumed, it can be assumed that there are no such transfers over the period under consideration. This means that fixed fee revenues and costs related to fixed voice access can be largely ignored. Regarding the usage based prices as in Armstrong's basic model (1998) linear retail tariffs (i.e. per minute prices) are considered.

To construct a comparable linear usage-based charge a homogeneous composite good of national voice traffic with a single per minute price for calls in all national networks is constructed. In reality, tariffs are as shown varying according to different parameters (e.g. day/night, call set-up fees, bucket plans, fixed/mobile network termination, etc). As fixed termination rates are assumed to be the same for all operators and to be (near) zero it is assumed that there are no cost-based reasons for discrimination of prices between different fixed networks. Given the large number of (other) dimensions of usage-based retail prices, any empirical model needs to make some form of tariff aggregation in order to ensure comparability. A possible way to calculate a single average per minute national tariff for fixed telephony is to divide all usage based national call revenues by the number of national traffic minutes (average revenue per minute (ARPM)). Doing so aggregates all dimensions explained earlier.

It should be noted that while this paper concentrates on traffic one aspect of access is taken into account which is that the number of accesses increases the extent of the network and potential users called. In this sense only (and exogeneously), it is expected that accesses influence traffic.

Given the above, it can therefore be concluded that the following assumptions are reasonable:

i)   fixed voice access fees are not considered
ii)   a single national per minute retail fixed voice calling price is considered

Regulation

In this chapter the regulation of fixed telephony operators in Europe and in particular in Switzerland is briefly reviewed. In the European Union, there is a clear trend towards deregulation of retail telephony markets as the European Commission stopped considering this market in need of ex-ante regulation in 2007[25]. Most member states have consequently started to withdraw regulatory remedies in this market in the years following the recommendation. According to Cullen, in 2013, six out of twenty-seven EU member states still regulated their telephony retail markets (Belgium, Bulgaria, Cyprus, Hungary, Poland, Portugal). Most regulations include some form of price control. Some national regulatory authorities (NRAs) have also demonstrated recently that competitive problems in this market persist, most recently Bulgaria in 2013[26]. The EU Commission has accepted such analyses' in several cases but advised the national regulatory authorities to reassess the situation in the next round of market analysis, as wholesale remedies may become sufficient to ensure retail competition in this market (e.g. case BG/2013/1421). In Switzerland, the regulatory framework has never foreseen the possibility of traditional asymmetric regulation of operators in the retail market. As has been described earlier, however, national retail price caps are in force for the universal service operator (incumbent) and seem to be binding (both for telephony access as well as fixed voice calls) for a standalone standard offer. There are, however, a number of subscriptions offered which imply that the prices charged by Swisscom may be substantially lower. Overall, the question of competition in retail markets for telephony services (national calls) seems in any case to be controversial across Europe.

Unlike the retail market, wholesale markets have been subject to active regulation not only in the rest of Europe but also in Switzerland. In particular, LRIC cost-based access prices for fixed call origination (since 1998) and for unbundled access to the local loop (only since 2008) are set by a regulator. However, regulation applies, unlike in the EU, only if parties cannot agree on terms of contract and there is a formal complaint to the Swiss Communications Commission (litigation based regime for regulation). Decisions of the authority may then be appealed at the Federal Administrative Court. Only after such a final decision regulated wholesale prices become binding which may make wholesale regulation in Switzerland slow in the sense that the market impact may be effective only years after the period under review. For example the fixed voice origination charges for the years 2000-2003 were lowered by 30% only in late 2006 upon regulatory intervention.

Market structure

In order to choose an appropriate market model the broad structure of the fixed voice telephony market needs to be analysed. It has to be pointed out that on top of the physical access line (copper, cable, FTTH), different technologies can be used to provide phone calls. While this may include traditional (PSTN and

---

[25] Commission recommendation on relevant product and service markets within the electronic communications sector susceptible to ex ante regulation - C(2007) 5406
[26] Case BG/2013/1421

ISDN) technologies, this may also include proprietary Voice over IP (VoIP) solutions (typically used by competitors over cable or DSL) as well as Voice over broadband (VoBB), where calls are made using a technology via the IP layer and "over the top" of a retail broadband connection (e.g. Skype) with a platform operated over the Internet. The latter technology is typically unable to guarantee quality of service. It is assumed in this paper that VoBB cannot offer sufficient quality of service to be considered a valid substitute. This choice is also necessary as no reliable data is available.

At the industry level, it can be seen that the number of active fixed voice accesses in Switzerland is slowly declining (Figure 1). This is not true for all types of accesses. While traditional PSTN/ISDN based telephone accesses are slowly declining both for the incumbent as well as for retailers (buying corresponding wholesale origination solutions from the incumbent), there is a steady increase of the number of proprietary VoIP accesses based on DSL or Cable. This increase is related to the market entry of Cable operators in 2004 as well as to the unbundling activities of competitors (in case of unbundling a fully independent VoIP telephony platform can be operated). This overall industry decline might be compensated to some extent by VoBB fixed telephone accesses offered by operators such as Skype.

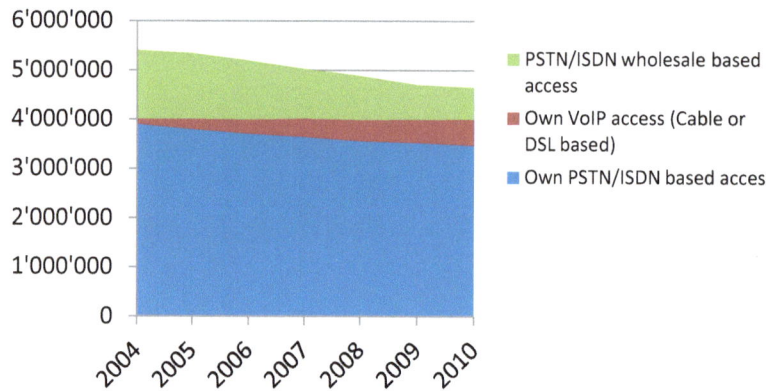

Figure 1 – Number of voice accesses per technology in Switzerland, Source: BAKOM

When looking at fixed voice subscribers per operator (Figure 2) for any technology excluding VoBB, it can be seen that the incumbent continues to have a very high and relatively stable market share of around 65 to 68% from 2007 to 2010, the central period of the period under review. In terms of traffic, it can be seen that Swisscom's market share is consistently about five percentage points lower than in the case of subscribers - at around 60 to 63% (see Figure 3). From this comparison, it seems that some firms (e.g. Sunrise) have customers with longer call duration.

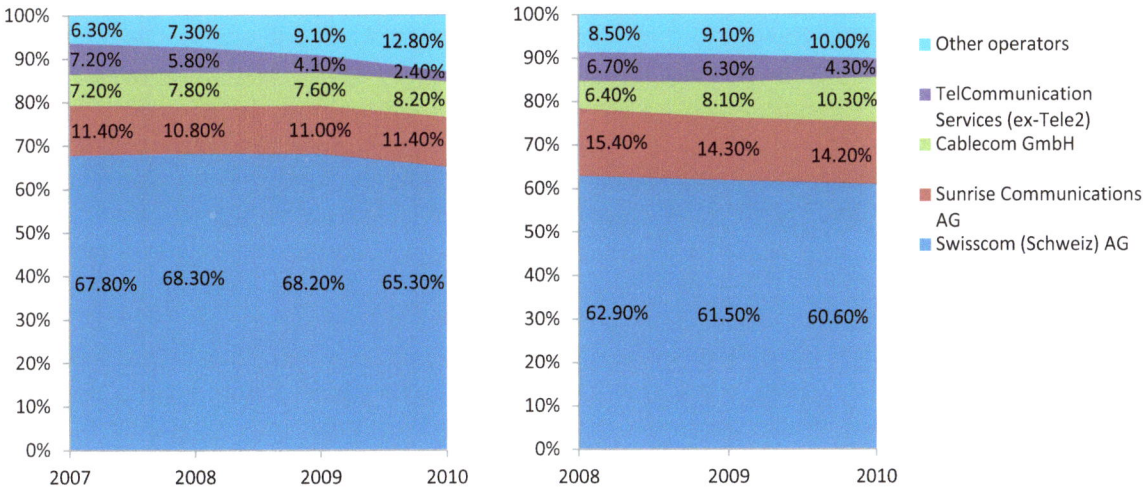

Figure 2 - Fixed voice subscriber market shares in Switzerland including VoIP, excluding VoBB, Source: BAKOM

Figure 3 - Fixed voice total traffic markets shares in Switzerland including VoIP, excluding VoBB, Source: BAKOM

In a detailed analysis of the market structure, the traffic data seems to indicate that next to the incumbent having a very high market share, many but significantly weaker competitors exist. The next biggest competitor, Sunrise (copper/DSL based), only holds 14% of subscribers at the end of 2010 (18% of traffic)[27]. Swisscom is therefore about three to four times larger than the next biggest competitor. Moreover, Cablecom (cable based) held 8% of subscribers and 10% of traffic in 2010. The remaining 10% of traffic are distributed among a large number of small competitors of which none exceeded 2% of subscribers or traffic (in total 53 operators were registered Swiss fixed voice operators).

Of the operators mentioned Sunrise has used only wholesale origination products from the incumbent (carrier selection and pre-selection) until 2008. From 2008, when a regulated unbundling offer was introduced the operator has increasingly migrated to unbundled products and its own VoIP based solutions. Cablecom in turn introduced its cable-based VoIP technology in 2004 (before the operator has not offered telephony services). The other smaller operators may use any of these ways to provide fixed voice services.

3. Theoretical framework

This chapter describes a market model providing the necessary structure for the estimation of (residual) demand elasticities and the degree of competition in the Swiss fixed telephony retail market.

The strong assumptions made in the preceding chapters allow for the design of a simple market model taking into account the volume of national fixed voice traffic (minutes) and corresponding linear retail prices (average revenue per minute). Access charges are assumed to correspond to efficient marginal cost based access prices set by the regulator, implying that both access-seeking operators as well as (assumed efficient) infrastructure-based operators can be assumed to face the same marginal network costs and compete on the same grounds. While mobile termination rates need to be considered as marginal cost

---

[27] This includes former independent competitor Tele2, merged with Sunrise in September 2008.

drivers, the assumed absence of fixed-mobile substitution and the independence of fixed operators from their potential mobile arms means that fixed voice services and operators are otherwise independent from mobile services and operators (i.e. mobile termination rates are assumed to be exogenous cost drivers). In addition, the relatively low level of regulated fixed termination charges when compared to mobile termination charges mean that when considering a single price for national calls, fixed termination costs and revenues can be ignored. Overall, this setting implies that a market model can be reasonably designed for the Swiss fixed telephony market, which is not significantly more complex than models which would be considered in markets without one- and two-way access aspects. The described simplifications allow for the estimation of a model even with the limited dataset available (2004-2012) described in the next chapter.

The basic framework used in this paper is known as the "dominant firm - competitive fringe" model, and was first proposed by Forchheimer (1908). While this model has relatively strong assumptions, they seem realistic in this specific market. Most importantly, competitors are fragmented and the relative size of the Swiss incumbent fixed telephony operator (60-65%) is clearly above 40%, the threshold for validity of the model indicated by Scherer and Ross (1990). Such a framework is, even under relaxed assumptions, shown in the next chapters to allow the estimation of the residual demand parameters of the incumbent as well as a range for the related conduct parameter measuring the degree of competition. The model is illustrated in Figure 4.

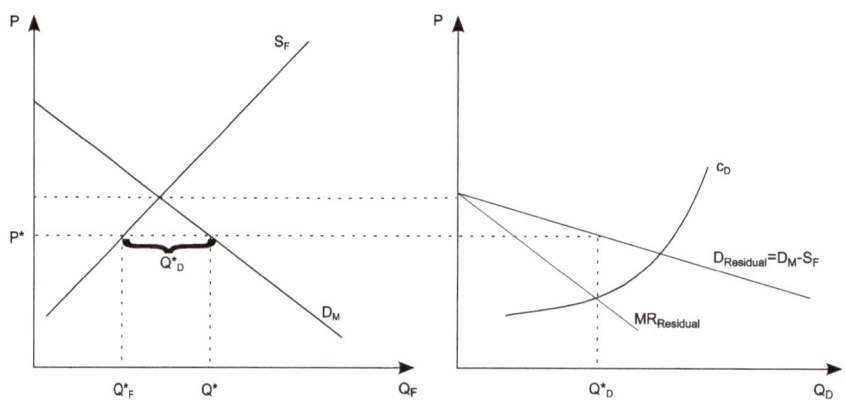

Figure 4 - The dominant firm - competitive fringe model.

Where:

D_M	is the market demand curve (Q is total demand)
S_F	is the fringe supply curve (Q_F is fringe supply)
$D_{Residual}$	is the residual demand curve the dominant firm faces (Q_D is dominant firm demand)
$MR_{Residual}$	is the residual marginal revenue curve the dominant firm faces
C_D	is the dominant firm's marginal costs
P	is the unique market price

In a dominant firm - competitive fringe model, the dominant firm takes into account market demand and fringe supply, and therefore finally the residual demand curve, maximising its profit by equating the resulting perceived marginal revenue to its marginal costs. The residual demand curve faced by Swisscom corresponds to the market demand curve minus the collective supply curve of the supposedly price taking fringe firms. As shown before, it is assumed that there is a largely independent (from mobile) and homogeneous good of national fixed voice traffic (calls to fixed and to mobile networks). A composite unique market price per minute is considered (average revenue per minute).

The dominant firm - competitive fringe model provides the basic structure necessary for the estimation of the residual demand curve and market conduct of the potentially dominant firm. Davis and Garcés (2009) describe a more general setting, with differentiated goods and potentially non-competitive fringe firms going back to Baker and Bresnahan (1988)[28]. Methodologically, the instrumental variable technique with firm specific cost shifters is used to estimate residual demand. This paper proposes a more simplified setting with a homogenous good and a competitive fringe. Practical implementations of the simplified dominant firm – competitive fringe model include Suslow (1986) in the aluminium industry and, more relevant for the present paper, Kahai, Kaserman and Mayo (1996) in the telecommunications industry, where market power of the incumbent telephone operator in the U.S. (AT&T) from 1984 to 1993 is estimated under a dominant firm hypothesis.

In the competitive fringe model it is assumed that the fringe firms are price takers and will – non-strategically - adjust their quantity to the given market price in order to maximize profits. Fringe supply therefore corresponds to the sum of marginal cost curves of fringe firms. The Lerner index measuring market power for the fringe firms is consequently equal to zero (see Lerner (1934)). The dominant firm has a competitive advantage in such a setting; it moves first and takes into account (with perfect foresight) the fringe's reaction and also market demand. In order to estimate the incumbent's residual demand elasticity and its market conduct, the residual demand function has to be estimated. First, in equation (1) market demand is considered.

$$Q = f(P,X) \tag{1}$$

Here, Q is measured by the total number of fixed telephony minutes demanded in the Swiss market (fixed-to-fixed as well as fixed-to-mobile calls), P is the average fixed telephony (traffic) price per minute (estimation of average revenue per minute considering all usage based revenues from fixed-to-fixed as well as fixed-to-mobile calls). X' is a vector of demand shifting variables. The fringe's marginal cost curve (horizontal sum of single fringe firm's marginal costs[29]) can be modelled in the following form.

$$C_F = f(Q_F, W_F, W_C)$$

Marginal costs are assumed to depend on the fringe's output level and to vary with a series of marginal cost shifters (some specific to the fringe, W_F, and others common to the industry, W_C). The fringe's marginal costs are in the segment under consideration assumed to increase with output due inefficiencies. This, as for example customer care cannot always be easily scaled and as there are capacity constraints of various network elements necessary for the provision of fixed telephony services. The fringe firm's first order condition for quantity choice is $P = C_F$, determining their cumulated supply curve (2).

$$Q_F = f(P, W_F, W_C) \tag{2}$$

The dominant firm, instead, chooses its profit maximizing market price, having consequently monopoly power over residual demand. The first order condition of the dominant firm (in this case it is equivalent for the firm to choose a profit maximizing price or quantity) lead to the following dominant firm supply relation determined by its marginal costs and perceived marginal revenue.

---

[28] For homogeneous goods see also Scheffman and Spiller (1987)
[29] When two identical firms can produce K at a given marginal cost, then both firms jointly can produce $2K$.

$$P(Q_D) + \frac{P(Q_D)}{\phi_1} = C_D(Q_D, W_D, W_C)$$

Here, ϕ_1, is residual demand elasticity - not to be confused with market demand elasticity - and C_D is the marginal cost function of the dominant firm. W_D represents marginal cost shifters specific to the dominant firm. As in this paper an isoelastic residual demand function[30] will be specified, this implies that the dominant firm supply function can be rearranged to the following expression:

$$P(Q_D) = \frac{C_D(Q_D, W_D, W_C)}{\left[1 + \frac{1}{\phi_1}\right]}$$

Finally, dominant firm's supply can be expressed in log-linear form (lower case variables for natural logs of variables) in (3)

$$p(q_D) = f_D^S(q_D, w_D, w_C) \tag{3}$$

In this case, the constant residual demand price elasticity becomes part of the constant of the function.

As usual, the dominant firm acts as a monopoly on its residual demand, meaning that (in contrast to the competitive fringe firms) it can influence the market price. The more price elastic residual demand is, the lower are prices are and the higher is output.

In this context the Lerner index describes the extent to which the dominant firm can raise prices above its marginal costs:

$$L = \frac{P - C_D}{P} = \frac{1}{\phi_1}$$

Similarly, dominant firm demand, Q_D, can be derived. The dominant firm takes into account market demand (1) as well as fringe supply (2) anticipating perfectly the fringes reactions.

$$Q_D(P) = Q(P, X) - Q_F(P, W_F, W_C)$$

Assuming that residual demand can also be expressed in log-linear form, it can be restated as (4)(see also Scheffman and Spiller (1987), Baker and Bresnahan (1988) and Davis and Garcés (2009)). Such a specification is beneficial in the empirical estimation because coefficients correspond to elasticities and allow for a straightforward discussion regarding percentage changes between variables.

$$q_D = f_D^D(p, x, w_F, w_C) \tag{4}$$

From residual demand (4) the original market demand and fringe supply parameters can usually not be calculated as combined effects are estimated. However, the system of dominant firm demand (4) and supply (3) can now be estimated using standard econometric tools for simultaneous equations. The price can be as usual instrumented using all exogenous variables in the system and be used to estimate the residual demand equation (2SLS). This technique exploits exogenous variation in price allowing for valid regression results. If a technique such as instrumental variables is not used, the assumptions of ordinary least squares regression are violated since the error is then correlated with price.

---

[30] i.e. with constant elasticity over the whole range of quantities considered

Before specifying the equations for estimation in detail, it should be noted here that the assumption of a dominant firm with monopoly power is not necessary for the described residual demand approach to hold. In fact, the incumbent can have more competitive conduct than that of a monopolist facing its residual demand and, in the extreme case, even behave as competitively as a price-taker and therefore like the fringe firms. As explained in Bresnahan (1982), different strategic conduct by the incumbent can be nested in its profit maximizing condition by simply adding a *conduct parameter, λ,* in the marginal revenue function (see also Davis and Garcés (2009). The approach to take the strategic conduct of firms as a parameter to be estimated and not as a model assumption is the core of the so-called "new empirical industrial organization" described in Bresnahan (1989). Including a conduct parameter λ for the potentially dominant firm, the above optimality conditions can be restated as

$$P(Q_D) + \lambda \frac{P(Q_D)}{\phi_1} = C_D(Q_D)$$

$$P(Q_D) = \frac{C_D(Q_D, W_D, W_C)}{\left[1 + \frac{\lambda}{\phi_1}\right]}$$

Here λ takes values between zero and one. $\lambda = 1$ corresponds to the working hypothesis of this paper of a dominant firm. In this case the incumbent behaves as a monopolist facing its residual demand. This setting is relaxed with $\lambda < 1$. In the extreme case, $\lambda = 0$. Then, the incumbent would be a price-taker without market power, as all its competitors. Any value between zero and one would correspond to the level of market power exercised (adjusted by elasticity). In particular, the Lerner index can be adapted to the following equation:

$$L = \frac{P - C_D}{P} = \frac{\lambda}{\phi_1}$$

As λ may be considered as a given parameter for the incumbent over the time horizon under review, the earlier derived incumbent's supply and residual demand functions remain valid if the incumbent does not behave fully as a dominant firm. The same is true for the instrumentation.

It should be noted that Bresnahan (1982) shows how to estimate λ in linear demand and marginal cost settings by using demand (and cost) rotating instead of shifting variables. In simpler settings, as in the context of this paper, λ can often not be directly estimated. For some particular values of the residual demand price elasticities, however, inferences on λ are possible. In particular, when residual demand is inelastic a maximal λ can be estimated for which the incumbents' action are compatible with profit maximizing behavior (i.e. perceived marginal revenue being positive). Concretely, it is shown in the next chapters that in light of the estimation results obtained (inelastic residual demand at -0.12), λ must be below one (and in particular below 0.12) to be compatible with profit maximization. The incumbent does therefore not act as a (purely) "dominant" firm. Nevertheless, the assumption of the incumbent acting as a dominant firm is for now maintained as a working hypothesis[31].

---

[31] It should be noted that the model remains also broadly valid when the assumption of a competitive fringe is relaxed (see Bresnahan and Baker (1988)).

First model equation (instrumented price):

As has been shown, considering all exogenous variables in the system, the market price can be instrumented. The system includes market demand, fringe supply and dominant firm supply. This means that demand shifters, common marginal cost shifters, fringe specific marginal cost shifters and, most importantly, dominant firm specific cost shifters need to be used as instruments to fit a variable $\hat{p}$ in (5).

$$\hat{p} = \theta_0 + w_D' \theta_D + w_F' \theta_F + w_C' \theta_C + x' \theta_B + \varepsilon \tag{5}$$

Second model equation (dominant firm residual demand)

The instrumented price can then be used to estimate the residual demand function of the (potentially) dominant firm (6).

$$q_D = \phi_0 + \phi_1 \hat{p} + w_F' \phi_F + w_C' \phi_C + x' \phi_B + \varepsilon \tag{6}$$

The coefficients of this function are estimated in the next chapter and allow inferences to be made on residual demand elasticity and market power. As has been shown before, the residual demand (price) elasticity is the result of a combination of market demand and fringe supply effects (see to (4)). Regarding the original individual coefficients of market demand and fringe supply, these can in general not be derived from the residual demand estimates in this context.

Some papers estimate fringe supply and market demand exclusively and make inferences on residual demand elasticity without direct estimation of the residual demand function. Kahai, Kaserman and Mayo (1996) use this approach as they argue that there are no sufficiently strong cost shifters for the dominant firm in order to estimate residual demand. While this may formally be a correct approach, the model applied by these authors largely abstracts from the dynamics of the most important factor of competition in this model; the potentially dominant firm. In the present paper, the marginal cost shifters of the incumbent are assumed to be important and are taken into account as instruments (e.g. staff numbers of Swisscom). Not taking these variables into account could cause a bias in the model independently of how it is specified. The model proposed here therefore allows for convenient direct estimation of the residual demand function and corresponding elasticities.

4. Input data

This chapter describes the dataset used to estimate the market model.

In the dataset considered in this paper, different quarterly time series to model competition in the fixed voice market in Switzerland from 2004 to 2012 are used. The strategic variables include the dominant firm's traffic output, i.e. Swisscom's fixed national outgoing calling minutes (to fixed and to mobile) and an estimate of market prices (average revenue per outgoing minute (subscription revenues excluded). The total market includes all fixed telephony technologies (e.g. PSTN as well as proprietary VoIP), but excludes web-based Voice over IP, as there is no reliable data on usage and prices and substitutability is assumed to be yet limited.

Traffic demand shifters include real GDP per capita, the network size (number of active fixed telephone lines in the industry, assumed to increases the number of potential calls) and time dummies. As it is assumed here that access is largely independent from traffic, the total network size is assumed to be exogenous, as in Kahai, Kaserman and Mayo (1996). Moreover, a series of common industry marginal cost shifters are considered, which include most importantly the average mobile termination rates, the actual regulated origination rates (as a proxy for a part of the network cost), interest rates and exchange rates (as a proxy of capacity investment costs). In addition, fringe cost shifters include the extent of usage of ULL and ADSL wholesale broadband products by alternative operators (essentially for a question of economies of scope reducing for example per unit customer care cost). Finally, cost shifters specific to Swisscom are considered, which include the number of staff (to control for possible remaining inefficiencies from times when the firm was fully state controlled). In addition, the number of ADSL lines sold by the incumbent may lead to economies of scope in a similar way as for the alternative operators. The effects described here are analysed in more detail in a later chapter interpreting the estimation results.

For the variables discussed, quarterly observations are available from Q4 2004 to Q2 2012. The variables in Table 1 are used to model the Swiss fixed voice market in the next chapter.

Variable	Definition	Name in Stata dataset	Unit	Source	N	Mean (abs)	Std. dev. (abs)
p	Price; average revenue per outgoing minute (subscription revenues excluded), deflated by CPI (100=2006), traditional and proprietary VoIP	*tradvoipallar pmallr*	CHF	Analysys Mason, BFS	55	0.098	0.017
q_D	Swisscom's fixed national outgoing calling minutes (to fixed and to mobile)	*tradvoipallmi nscm*	Minutes	Swisscom	55	2.45E+09	5.24E+08

Demand shifters

Variable	Definition	Name in Stata dataset	Unit	Source	N	Mean (abs)	Std. dev. (abs)
x_1	Income; real GDP per capita (100=2006)	*yrealcapita*		Seco / BFS	51	0.013	0.0006
x_2	Number of active telephony lines (PSTN)	*pstn*	000s	Swisscom	56	2'693	1'071
x_3, x_3, x_4	Quarterly dummies for quarters 2, 3 and 4	*d2, d3, d4*					

Fringe supply shifters

Variable	Definition	Name in Stata dataset	Unit	Source	N	Mean (abs)	Std. dev. (abs)
w_{F1}	Number of wholesale ADSL lines sold by Swisscom	*adslwhole*	000s	Swisscom	47	262.996	147.691
w_{F2}	Number of ULL access lines sold by Swisscom	*ullreal*	000s	Swisscom	59	53.288	103.362

Common supply shifters

Variable	Definition	Name in Stata dataset	Unit	Source	N	Mean (abs)	Std. dev. (abs)
w_{C1}	Average regional fixed voice origination prices per minute deflated by CPI (100=2006)	*scmregorr comcregorr*	CHF (100= 1/2000)	Swisscom /ComCom	55	0.012	0.004
w_{C2}	Weighted average of mobile termination rates deflated by CPI (100=2006)	*wavgmtrr*	CHF	Operators Analysys Mason, BFS	34	0.21	0.089
w_{C3}	Interest rates on 30y bonds of the Swiss Confederation	*interest30y*	%	Swiss national bank	56	3.19	1.026
w_{C4}	Exchange rate EUR/CHF	*fxratechfeur*	EUR/C HF		56	0.676	0.067

Dominant firm supply shifters

Variable	Definition	Name in Stata dataset	Unit	Source	N	Mean (abs)	Std. dev. (abs)
w_{D1}	Number of staff working for Swisscom (Group)	*SCMstaff*	units	Swisscom	35	18418	1880.025
w_{D2}	Number of active retail ADSL lines (Swisscom)	*adslretail*	000s	Swisscom	47	885	602

Table 1 – Input data from the Swiss fixed telephony market (2004-2012)

It should be noted that for the estimation natural logs of all variables are taken (the Stata variables names in the technical annex in this case have an "ln" prefix[32].

5. Econometric analysis and estimation

This chapter reports the estimation results. As is typical in a demand and supply setting, instrumental variables can be used for estimation. The basic econometric model can be analysed in two stages as described in the analytical framework. In the technical annex (Section 9.1.) a "baseline" or reference model in line with the analytical framework is described and estimated. It is shown that important adjustments to the basic model are necessary in order to resolve the detected econometric problems. Most importantly, it is found that fixed voice traffic and prices are decreasing over time, implying non-stationarity both for the incumbent's and the fringe's strategic variables. In addition, errors are serially correlated and strategic variables seem to be autocorrelated (most importantly with their first lag). The technical annex analyses these problems and develops the solution adopted in this section in the form of an Auto-Regressive Distributed Lags (ARDL[33]) model. It specifically implies that the baseline equations are added one period lagged dependent and independent variables (ARDL (1,1)). This adaption is necessary to exclude spurious regression results as a consequence of non-stationarity and insufficient cointegration of variables. As the model becomes dynamic, the interpretation of specific coefficients becomes more complex. While coefficients continue to correspond to short run (same period) direct effects on the dependent variable ("impact multipliers"), the long term effects of a (permanent) change in an explanatory variable need to be calculated ("long run multipliers", see Equation (9) in Section 9.7. of the technical annex).

The technical annex overall concludes that a first stage ARDL (1,1) regression is estimating an instrumented variable with instruments (all exogenous variables in the system) that are non-stationary, but having a sufficient degree of cointegration. Similarly, the second stage ARDL (1,1) regression is estimated with variables that are non-stationary, but having a sufficient degree of cointegration (including the instrumented variable). In addition, in both regressions with these specifications, there is no serial correlation of errors anymore. Overall, both the first and second stage ARDL regressions are valid. Hence, in the next section, the ARDL (1,1) regressions are used to correct the baseline model (see technical annex) for the econometric problems identified, especially spurious regression.

Estimation

All necessary tests to exclude spurious regression results have been performed in the technical annex. The following ARDL (1,1) second stage regression (following a similar first stage regression) should therefore represent the optimal 2SLS model in the context of this paper and provide valid estimation results for the incumbent's residual demand function. For convenience, Equation (12) from the technical annex specifying residual demand of the incumbent may be restated here:

$$q_{D,t} = \hat{\phi}_0 + \hat{\phi}_{1,t}\hat{p}_t + \hat{\phi}_{1,t-1}p_{t-1} + \hat{\phi}_{2,t-1}q_{t-1} + w'_{F,t}\hat{\phi}_{F,t} + w'_{F,t-1}\hat{\phi}_{F,t-1} + w'_{C,t}\hat{\phi}_{C,t} + w'_{C,t-1}\hat{\phi}_{C,t-1} + x_t'\hat{\phi}_{B,t} + x_{t-1}'\hat{\phi}_{B,t-1} + \epsilon_t \tag{12}$$

Estimation results are reported in Table 2, where the coefficients (impact multipliers) of the ARDL regression are represented next to the baseline model coefficient estimates. The latter are, however, as

---

[32] Having only about 30 observations, it is early to estimate this market model. A later estimation could take advantage of more data points. Standard errors seem to be high, which is often the case in small samples.
[33] Or also ADL

shown in the annex, likely to be spurious. The dynamic multipliers (impact on following period) are represented one row below (L1). Finally, another row below long run multipliers, $\hat{\phi}^{LR}$, are calculated, which represent the effect of a change in an independent variable over the whole time horizon on the dependent variable. The last column compares the coefficient estimates of the ARDL(1,1) estimation with the baseline estimation. It should be noted, however, that from the tests conducted the ARDL model is clearly the correct model for estimation and the baseline model results are only reported for convenience.

Para-meters	Variable	ARDL (1,1) estimates			Baseline estimates			Comparison
		Coefficient estimate	Robust Std. err.	P>|t|	Coefficient estimate	Robust Std. err.	P>|t|	Difference ADLR/Baseline (to impact and LR multiplier) in %
$\hat{\phi}$	q_D							
$\hat{\phi}_{2,t-1}$	L1	-0.181	0.100	0.069				
$\hat{\phi}_{1,t}$	p	-0.044	0.044	0.321	-0.663	0.364	0.068	-93%
$\hat{\phi}_{1,t-1}$	L1	-0.102	0.049	0.038				
$\hat{\phi}_1^{LR}$		-0.124						-81%
$\hat{\phi}_{B1,t}$	GDP per capita	1.364	0.302	0.000	-0.333	0.726	0.646	-509%
$\hat{\phi}_{B1,t-1}$	L1	0.387	0.236	0.102				
$\hat{\phi}_{B1}^{LR}$		1.482						-545%
$\hat{\phi}_{B2,t}$	PSTN lines	0.876	0.159	0.000	1.979	0.366	0.000	-56%
$\hat{\phi}_{B2,t-1}$	L1	1.912	0.337	0.000				
$\hat{\phi}_{B2}^{LR}$		2.360						19%
$\hat{\phi}_{B3,t}$	d2	-0.030	0.005	0.000	-0.032	0.016	0.042	-5%
$\hat{\phi}_{B4,t}$	d3	-0.098	0.006	0.000	-0.065	0.024	0.007	51%
$\hat{\phi}_{B5,t}$	d4	-0.025	0.010	0.008	-0.021	0.013	0.112	23%
$\hat{\phi}_{F1,t}$	ADSL wholesale lines	-0.233	0.058	0.000	-0.220	0.043	0.000	6%
$\hat{\phi}_{F1,t-1}$	L1	0.046	0.054	0.391				
$\hat{\phi}_{F1}^{LR}$		-0.158						-28%
$\hat{\phi}_{F2,t}$	ULL lines	-0.003	0.001	0.011	-0.001	0.001	0.548	315%
$\hat{\phi}_{F2,t-1}$	L1	0.000	0.001	0.700				
$\hat{\phi}_{F2}^{LR}$		-0.002						196%
$\hat{\phi}_{C1,t}$	Origination prices	-0.097	0.026	0.000	0.150	0.133	0.259	-165%
$\hat{\phi}_{C1,t-1}$	L1	0.037	0.026	0.156				
$\hat{\phi}_{C1}^{LR}$		-0.051						-134%
$\hat{\phi}_{C2,t}$	MTR	0.010	0.021	0.621	0.116	0.071	0.104	-91%
$\hat{\phi}_{C2,t-1}$	L1	0.157	0.018	0.000				
$\hat{\phi}_{C2}^{LR}$		0.141						22%
$\hat{\phi}_{C3,t}$	Interest rate	-0.112	0.034	0.001	-0.050	0.052	0.338	123%
$\hat{\phi}_{C3,t-1}$	L1	0.070	0.025	0.005				
$\hat{\phi}_{C3}^{LR}$		-0.035						-30%
$\hat{\phi}_{C4,t}$	Exchange rate	0.197	0.141	0.163	-0.220	0.155	0.155	-189%
$\hat{\phi}_{C4,t-1}$	L1	0.362	0.149	0.015				
$\hat{\phi}_{C4}^{LR}$		0.473						-315%
$\hat{\phi}_0$	_cons	11.038	1.928	0.000	4.390	4.102	0.285	151%
$\hat{\phi}_0^{LR}$		9.343						113%
		R^2	0.998		R^2	0.966		
		Prob> χ^2	0.000		Prob>F	0.000		
		N	27		N	28		

Table 2 – Estimation results, Baseline and ARDL(1,1)[34]

As has been shown, the results of the ARDL(1,1) regression should not be the result of a spurious regression and can be interpreted as usual in the next section.

6. Interpretation of the results

In this chapter the main estimation results are discussed. After performing a detailed analysis of the stability of the baseline results and introducing the necessary corrections to account for cointegration and serial correlation of errors (see technical annex), the resulting autoregressive distributed lags model regression results (Table 2) can be interpreted as usual 2SLS results. This section shows how, as for any model which includes intertemporal effects, the effect of a change in one variable on another variable has to be divided into an immediate same period effect (impact multiplier) and a long term effect (long run multiplier). It is possible to use the estimated coefficients to calculate functions illustrating the intertemporal adjustment behaviour of the dependent variable after exogenous shocks over time.

For convenience (9) from the technical annex can be restated here:

$$y_t = c + \alpha_1 y_{t-1} + \beta_0 z_t + \beta_1 z_{t-1} + \epsilon_t$$

From the explanations in the technical annex (see (9) and (10)) it follows that the time t+k multiplier (i.e. the coefficient explaining the impact of a marginal change of an independent variable in period t on the dependent variable from t until t+k) is described by the following expressions:

$$\mu_{t+k} = \frac{\partial y_t}{\partial z_t} + \frac{\partial y_{t+1}}{\partial z_t} + \frac{\partial y_{t+2}}{\partial z_t} + ..$$

$$= \beta_0 + (\alpha_1\beta_0 + \beta_1) + \alpha_1(\alpha_1\beta_0 + \beta_1) + \alpha_1^2(\alpha_1\beta_0 + \beta_1) + ...$$

If $k = 0$ $\quad\quad\quad \mu_{t+k} = \beta_0$

If $k \neq 0$ $\quad\quad\quad \mu_{t+k} = \beta_0 + \sum_{k=1}^{t} \alpha_1^{k-1}(\alpha_1\beta_0 + \beta_1)$

If $k \to \infty$ $\quad\quad\quad \mu_{t+\infty} = \sum_{k=0}^{\omega} \alpha_1^k(\beta_0 + \beta_1)$ $\quad\quad\quad\quad\quad\quad\quad\quad\quad$ (7)

The multiplier with k=0 is the called the impact multiplier.

As long as $\alpha_1 < 1$, which is assumed here[35], the long term multiplier defining the effect on the steady state of the dependent variable from a change in an independent variable in period t corresponds to (8) (Johnston and Di Nardo (1997)).

$$\frac{dy^*}{dz_t} = \frac{\beta_0 + \beta_1}{1 - \alpha_1}$$ $\quad\quad\quad\quad\quad\quad\quad\quad\quad\quad\quad\quad\quad\quad\quad\quad\quad$ (8)

---

[34] Stata command: ivregress 2sls lnradvoipallminscm l.lnradvoipallarpmallr l.lnradvoipallminscm d2 d3 d4 l(0/1).lnyrealcapita l(0/1).lnpstn l(0/1).lnscmregorr l(0/1).lnadslwhole l(0/1).lnullreal l(0/1).lnwavgmtrr l(0/1).lninterest30y l(0/1).lnfxratechfeur (lnradvoipallarpmallr=l(0/1).lnSCMstaff l(0/1).lnadslretail) , first robust
[35] otherwise effects on the dependent variable over time become ever stronger

These functions, representing the short, medium and long run multipliers, may in the present log-linear model also be interpreted as "cumulative" elasticities for a period up to k periods after the assumed shock in the independent variable of interest. Calculating these effects, Figure 5 represents the percentages of full adjustment after k quarters after the shock, where the full adjustment corresponds to the long run multiplier. This contrasts with the baseline regression, where it is assumed by definition that 100% of the adjustment is taking place in the same period as the shock. It can be seen that residual demand in the estimated model nearly fully adjusts to shocks in the market in all cases after only four quarters. A large part of the full adjustment for most shocks takes place in the same quarter as the shock as well as in the following quarter. It can also be noted here that for several variables there seems to be some form of overshooting effect in the quarter of the shock .

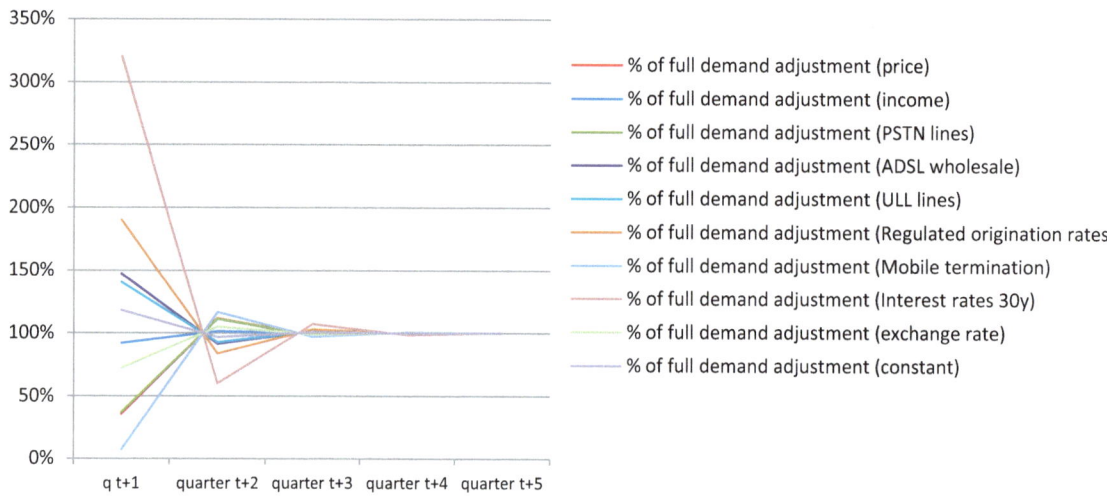

Figure 5 – Percentage of long term adjustments in the first quarters after the shock

Considering the above, the interpretation of results can be limited to effects in the first four quarters after the shock and most importantly the period of the shock and the following period. The most important estimated cumulative residual demand elasticities for the incumbent are reported in Figure 6.

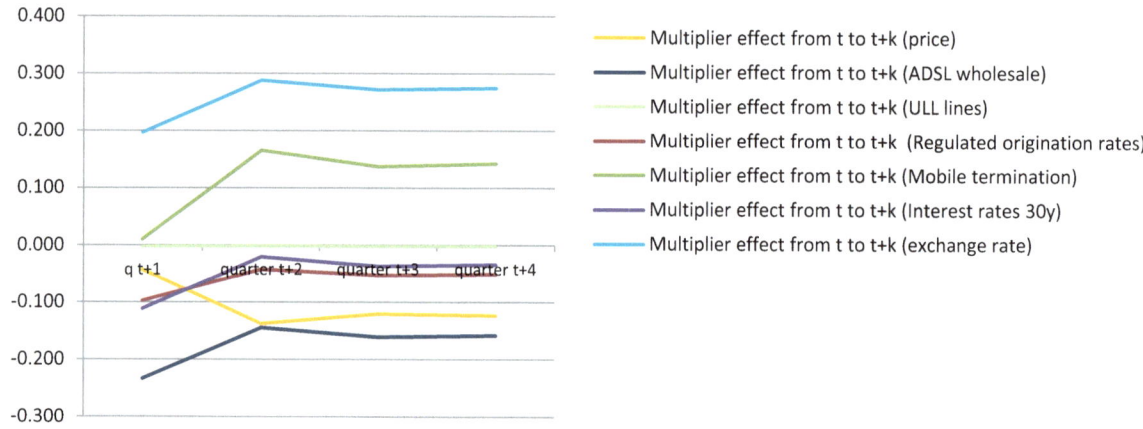

Figure 6 – Main estimated cumulative residual demand elasticites up to period t+k

In the following sections, the single coefficient estimation results and corresponding cumulative elasticities are discussed.

6.1. Price elasticity of residual demand

For convenience, single coefficient results of Table 2 are reported again in the corresponding subsections before interpreting the coefficients.

Variable		ARDL (1,1) estimates			Baseline estimates			Comparison
		Coefficient estimate	Robust Std. err.	P>|t|	Coefficient estimate	Robust Std. err.	P>|t|	%
p		-0.044	0.044	0.321	-0.663	0.364	0.068	-93%
	L1	-0.102	0.049	0.038				
	LR	-0.124						-81%

Table 2.1. – Estimation results, coefficient on price

It can be seen that the elasticity of residual demand of Swisscom in response to a price change is very low during the quarter in which prices are adjusted (-0.044). The direct effect on the period following the shock is higher, at -0.102. Finally, compared to the same period effect, the long term cumulative elasticity, taking into account all of the price change on residual demand over subsequent periods, triples to -0.124.

It may be seen as unusual in as fast paced an industry as telecommunications that a large part of the demand adjustment takes several months (or even quarters) to materialize. The contractual terms of Swisscom (and others operators) in the market during the period under review may, however, explain this phenomenon. A reduction in voice minutes demanded may come from users calling less or from users fully giving up fixed telephony services. In the period under review, Swisscom required that its private customers provide a 60 days' notice period to cancel the service. During the first 60 days after the adjustment, volume may therefore only be affected by customers calling less. Afterwards, it will also be affected by customers fully giving up their fixed access lines (or switching). This means that for two-thirds of a period after the shock a large part of the adjustment is blocked and possibly compensated only subsequently[36].

Nevertheless, even taking into account the long term effects of a price change for fixed voice traffic, residual demand of Swisscom seems to be highly inelastic (-0.124). This indicates that Swisscom, acting as a dominant firm, could increase prices without fearing immediate and even medium and long run material residual demand adjustments. In particular, after a 10% price increase, demand for Swisscom fixnet traffic would decline by only 0.4% in the same quarter, 1% in the following quarter (direct impact only) and 1.2% cumulated in the long term.

As has been shown earlier for any type of demand function, the marginal revenue of a dominant firm is given by

$$MR = P(Q)\left[1 + \frac{1}{\phi_1}\right]$$

Where ϕ_1 is estimated at -0.124. When this price elasticity is lower (in absolute terms) than 1, marginal revenue is by definition negative. As on the other hand marginal costs are always positive, a profit maximizing dominant firm would in such a situation increase prices and reduce its output up to a point

---

[36] It should be noted that, in the particular case of standard rates, upwards price adjustments are often notified in advance, implying that customer reactions in this case may be more immediate.

where residual demand becomes elastic and where MC=MR. This is not directly possible in the model proposed, as the log-linear form of residual demand assumes constant price elasticity. This is, however, a simplifying assumption (see Davis and Garcés (2009)). If the incumbent would actually increase prices, this would in reality possibly lead an increased price elasticity and the model would also find a higher (constant) level of price elasticity. In any case, the model and its estimates indicate that at this outcome Swisscom cannot be a profit maximizing dominant firm. Graphically, the estimated functions can be illustrated in Figure 7.

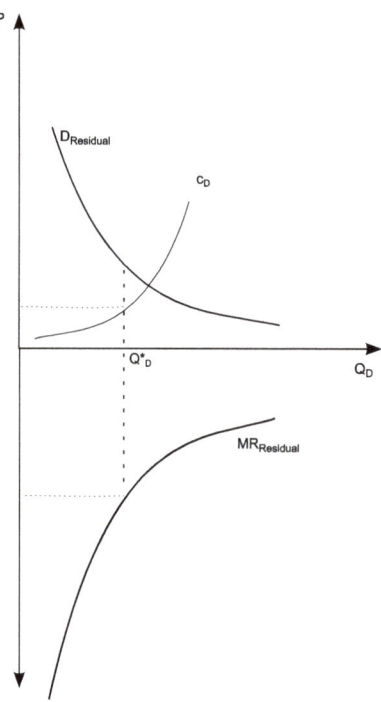

Figure 7 – Estimated residual demand curve and marginal revenue (dominant firm)

As also explained earlier, the market model is also compatible with the incumbent acting strategically differently, i.e. as a price taker (no market power) or between a price taker and a monopolist facing residual demand. The marginal revenue becomes:

$$MR = P(Q)\left[1 + \frac{\lambda}{\phi_1}\right]$$

For Swisscom to be able to behave in a profit maximizing manner, marginal revenue would need to be positive. With $\phi_1 = -.12$, this is the case only when $\lambda < 0.12$. In this case, the marginal revenue curve would as usual be positive and a profit maximizing firm could produce at the intersection point with marginal cost (see Figure 8).

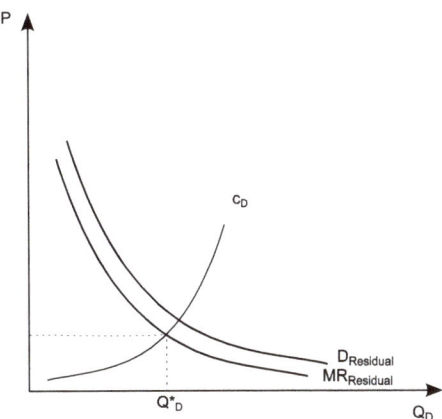

Figure 8 - Estimated residual demand curve and marginal revenue (not fully dominant firm; $\lambda < 0.12$)

This implies that only when Swisscom acts not as a dominant firm but nearly as competitively as the fringe firms, its actions are compatible with the standard assumption of profit maximization. It can therefore be concluded that Swisscom's conduct is largely competitive. To cite one example, Qu (2007) finds even lower price elasticity of residual demand, not significantly different to zero for strategic firms in the wholesale electricity market in the U.S. He similarly concludes that this implies that the incumbent's behaviour is consistent with fully competitive pricing. It should be noted that this is not the case in this paper. While the immediate, same period effect (-0.044) is also not significantly different than zero, subsequent effects are. The estimates in this paper imply only that Swisscom must act rather competitively. It can, however, not be affirmed whether or not the firm acts fully competitively.

It should also be noted, that for a small part of the considered retail telephony price basket, binding price regulation via universal service exists (fixed-to-fixed calls for standalone offers). For these prices, the fact that Swisscom is not increasing prices as a dominant firm while facing such inelastic demand may be not related to competition but to regulation. It can therefore not be said a priori for these prices that lifting regulation would not lead to increased prices, as currently regulated prices distort this analysis. These regulated prices are, however, only of limited validity as a large number of Swisscom price plans and subscriptions over the period under review foresee non standard per minute rates. For the composite good under analysis (the incumbent's fixed retail voice traffic overall) the conclusion should therefore hold that the incumbent behaves largely competitively and that any price regulation of this market is unnecessary and potentially distorting competition.

6.2. Demand shifters

Variable	ARDL (1,1) estimates			Baseline estimates			Comparison
	Coefficient estimate	Robust Std. err.	P>| t		Coefficient estimate	Robust Std. err.	P>| t		%
Income	1.364	0.302	0.000	-0.333	0.726	0.646	-509%
L1	0.387	0.236	0.102				
LR	1.482						-545%
PSTN lines	0.876	0.159	0.000	1.979	0.366	0.000	-56%
L1	1.912	0.337	0.000				
LR	2.360						19%

Income

According to the estimates an increase of per capita income of 10% is expected to correspond to an increase in fixed voice traffic demand for the incumbent of 14.8%. This effect may be slightly stronger than expected, with an income elasticity higher than one, meaning that fixed voice traffic in this context is suggested to be superior good.

While studies usually find that fixed voice traffic is a normal good (e.g. Ahn, Lee and Kim (2002) for Korean fixed telephony), there are also studies supporting the view of a superior good, as for example, Gyimah-Brempong and Karikari (2007) for African countries. Agüero, De Silva and Kang (2011) similarly review Engel curves for (essentially fixed) telecommunications services and find that early studies and studies on developing countries mostly found fixed telephony services to be a luxury good, while in developed countries these would usually be rather normal goods. McCloughan and Lyons (2006) review income effects on mobile telephony services which they find to be usually a normal or superior good. They also state that the income elasticity of demand may depend on the proportion of high income customers served. Vogelsang (2010) overall shares these views stating that mobile telephony may be a luxury good initially but may become a normal good in a more mature phase.

It may seem reasonable that demand for (more expensive) mobile calls increases more strongly with an increase in income than demand for fixed calls (the cheaper and less convenient alternative). Nevertheless, it also seems reasonable that both goods in the period under examination were perceived as superior goods in Switzerland, as unregulated mobile termination rates and telephony prices from any network towards mobile were as shown extraordinarily high.

PSTN lines

The model suggests also that there are positive network effects in the sense that the larger the fixed network (PSTN lines[37]) the larger the amount of persons that can be reached from fixed accesses via particularly cheap fixed-fixed calls suitable for longer calls. As also in Kahai, Kaserman and Mayo (1996) the total network size is assumed to be exogenous and therefore largely independent from traffic (standalone fixed access prices of all operators have for instance remained constant in the period under analysis and do not vary across operators). It should be noted here that the PSTN lines variable is also representing the effects of growth in population. This variable has actually been dropped due to strong colinearity with PSTN lines.

6.3.   **Fringe cost shifters**

Variable	ARDL (1,1) estimates			Baseline estimates			Comparison
	Coefficient estimate	Robust Std. err.	P>|t|	Coefficient estimate	Robust Std. err.	P>|t|	%
ULL lines	-0.003	0.001	0.011	-0.001	0.001	0.548	315%
L1	0.000	0.001	0.700				
LR	-0.002						196%

---

[37] VoIP lines were largely irrelevant for large part of the period under review.

ADSL wholesale lines	-0.233	0.058	0.000	-0.220	0.043	0.000	6%
L1	0.046	0.054	0.391				
LR	-0.158						-28%

Table 2.3. – Estimation results, coefficients on fringe cost shifters (ULL lines, ADSL wholesale lines)

The level of ADSL wholesale lines sold by the incumbent (to the fringe firms) and the level of usage of unbundling (ULL) are proxies for the efficiency with which fringe telephony operators can operate their central infrastructure and services (while regional origination costs are assumed to be the same for all operators). Efficiencies are mainly expected due to economies of scope by providing both telephony and Internet services.

The Internet market is assumed to be largely independent from the telephony market here, which is why the ADSL wholesale lines are taken as exogenous. The level of unbundling needs not necessarily be correlated with broadband penetration (see Nardotto, Valletti and Verboven (2012)). However, unbundling can in this market model be expected to mainly indicate the technological state of the fringe firms' backbone network, as for a large part of the period under review ULL was not available and for a subsequent relatively long period the ULL infrastructure was only successively rolled out starting with the most profitable areas. It can therefore be expected that the market environment (such as wholesale and retail prices) could have affected the level of demand for ULL only for a small part of the period under review.

The results show, however, that the efficiency effect by ULL in the backbone network seems to be limited. The number of (wholesale) broadband lines provided by the fringe firms seem, however, to have a more substantial cost reducing effect expanding as expected fringe supply and limiting residual demand.

6.4. **Common cost shifters**

Variable	ARDL (1,1) estimates			Baseline estimates			Comparison
	Coefficient estimate	Robust Std. err.	P>| t |	Coefficient estimate	Robust Std. err.	P>| t |	%
Mobile termination rates	0.010	0.021	0.621	0.116	0.071	0.104	-91%
L1	0.157	0.018	0.000				
LR	0.141						22%
Origination prices	-0.097	0.026	0.000	0.150	0.133	0.259	-165%
L1	0.037	0.026	0.156				
LR	-0.051						-134%
Interest rate	-0.112	0.034	0.001	-0.050	0.052	0.338	123%
L1	0.070	0.025	0.005				
LR	-0.035						-30%
Exchange rate	0.197	0.141	0.163	-0.220	0.155	0.155	-189%
L1	0.362	0.149	0.015				
LR	0.473						-315%

Table 2.4. – Estimation results, coefficients on common cost shifters (MTR, origination, interest and exchange rate)

The interpretation of common cost shifters has to be done carefully. A common positive cost shock affects both the fringe firms as well as the dominant firm. The above coefficients measure, however, only the impact via fringe supply on residual demand. The impact on incumbent supply is not measured and not the focus of this paper.

Mobile termination rates

The voice traffic considered includes fixed to mobile calls. In the Swiss setting, with unregulated (and relatively high) mobile termination rates in the period under review, these probably are the most important cost drivers for national traffic (fixed termination costs and revenues continue to be abstracted from here). As explained earlier, these rates are assumed to be exogenous. In such a case, it can be expected that any drop in the average mobile termination rate would have a positive effect on fringe output reducing dominant firm demand. It is found that an overall decrease of 10% in mobile termination rates would reduce incumbent demand by 1.4% in the long term. Again, incumbent equilibrium output may differ as the effect of these common costs on incumbent supply would also need to be considered.

Exchange rates

Telecommunications equipment is typically supplied by firms outside Switzerland. When the strength of the local currency increases it can be expected that telecoms equipment may be sourced at lower prices. The Euro area is by far the largest import area for the Swiss economy, the EUR/CHF is therefore considered. An increase in the exchange rate EUR/CHF (around 0.8 in 2012) would imply that more Euros are obtained per Swiss Franc. In this case the purchasing power of Swiss companies should at international level increase and sourcing prices decrease. This effect should be common for all firms in the market. The effect on fringe firms would be an expansion of supply and therefore a reduction in dominant firm demand. This effect is not seen in the results, however. In fact, the same period effect is insignificant, while the one quarter lagged effect is slightly positive. It is therefore likely that such equipment is not a relevant marginal cost driver in the fixed voice market.

Interest rates

Another common cost driver is capital costs proxied by the interest rate (Swiss 30-year Government Bond). An increase in capital costs is expected to lead to a reduction of fringe supply and an increase in the incumbents' residual demand. Here the lagged effect is positive as expected but the same period and the long term effects are negative. Again, in this market marginal costs may be only partially affected by significant capital investment.

Finally, it may be noted that the state-owned incumbent may be less dependent on the capital market. This is, however, a supply side consideration which is not directly considered here.

Origination rates

Incurred marginal costs for network origination are as explained earlier assumed to be the same for all operators (infrastructure-based operators as well as access-seekers). Regulated origination rates are used here as a proxy for (regional) network origination cost for all operators.

An increase of origination rates for providers for fixed voice calls should increase all operators' origination cost and reduce fringe output increasing Swisscom's residual demand. While this is again the case for the one period lagged effect, this is not the case for the same period as well as the long term effect. This ambiguous effect may mean that origination rates may be of limited importance on the retail market. This may be reasonable given their comparatively low level when compared to mobile termination rates (per minute).

7. Conclusions

The simple simultaneous equations framework proposed in this paper, based on a generalised "dominant firm – competitive fringe" model has allowed to estimate the Swiss incumbent's residual demand function for fixed telephony traffic for the period from 2004 to 2012. Unlike earlier papers, this paper directly estimates residual demand using dominant firm specific cost shifters and ensures a sufficient level of cointegration to avoid spurious regression results. Evidence of a competitive fixed telephony market in Switzerland during the period under review is found, calling into question the need for continued regulation.

While conduct cannot be directly estimated using the framework described, the concrete estimates show that demand is inelastic (long run price elasticity of -0.12). Such a level of elasticity is, however, only compatible with a profit maximising incumbent in the case of largely competitive conduct (conduct parameter below 0.12 and therefore close to zero). It is therefore found that Swisscom acted rather competitively in the fixed telephony retail market in the period under review. If the problem of an uncompetitive retail market ever existed, it seems that the entry of alternative operators using cable infrastructure and the introduction of regulated wholesale access (carrier pre-selection as well as local loop unbundling) have successfully removed it. This implies that the (partial) retail price caps in place in Switzerland can no longer be justified on the basis of a lack of competition and should be removed as has been done recently in a number of other European states. Similar conclusions can be drawn for possible universal service objectives. As regulated wholesale products (call origination) are available on national scale at a uniform price, retail competition should be ensured at national scale as well and not only locally.

The model in this paper is based on a series of strong assumptions. Future empirical work should try to relax them. Most importantly, the model assumes that all operators face the same marginal network cost. In reality, this may not be the case. The incumbent may face lower internal marginal network costs in some form. In this case, vertical integration of the incumbent becomes relevant. As has been discussed, Inderst and Peitz (2012) show then that in case of price dependent demand[38], the incumbent could also charge lower uniform retail prices in equilibrium than its competitors and that it then has a higher market share (partial foreclosure). Such pricing policy corresponds to a margin squeeze by the incumbent (see Vickers (2005)). If such a margin squeeze has occured, competitive conduct by the incumbent in the retail market could be overestimated. Different firm-specific marginal network costs and prices (differentiated goods) should therefore be considered. More generally, the market for inputs should be modelled in more detail in future work. This could be of particular interest, as the current regulatory debate on European level is starting to focus on these markets. In particular, an Ecorys study for the European Commission (2013) concludes that regulated unbundling and wholesale broadband access products sufficiently constrain the incumbent in its call origination pricing in the future, suggesting it may be possible to lift also the regulation of call origination[39]. It is possible that this paper's focus on this regulated access product was inappropriate, even retrospectively, and may have caused the unexpected sign of the related coefficient as discussed in the last chapter. Whether the European Commission shares the view of Ecorys remains to be seen. An updated recommendation on relevant markets is due in the second quarter 2014.

Lastly, fixed-mobile substitution is becoming increasingly important. Empirical work reviewing more recent periods should account for this by treating mobile telephony as a potential substitute. Furthermore, fixed termination rates are low when compared to mobile termination rates, but not zero. Such termination rates

---

[38] when each competitor has a price dependent hinterland of loyal customers unaccessible to the other operator.
[39] as well as wholesale line rental

may be above marginal costs, implying that termination costs and revenues need also to be considered in the profit maximising problem of the incumbent. As has been described earlier, this may imply waterbed effects where a decrease in wholesale prices may also lead to an increase in particular retail prices. In addition, an extended model could consider web-based VoIP as data becomes available as well as two-part tariffs. It should be noted, however, that the current limited availability of (quarterly) data points greatly restricts the complexity of an empirical model in this market. Any relaxation of the hypotheses described here would therefore not be implementable with the dataset used in this paper. While future work could reformulate the model and estimate it when more data points become available (or apply it to a market where more frequent data is available[40]), the present model represents a starting point, implementing the simplest possible specification with limited data.

---

[40] e.g. Austria, where monthly data is available

8. Bibliography

Agüero, A., de Silva, H., & Kang, J. (2011). Bottom of the Pyramid expenditure patterns on mobile services in selected emerging Asian countries. *Information Technologies & International Development, 7(3)*, pp-19.

Ahn, H., Lee, J., & Kim, Y. (2002). Estimation of a fixed-mobile substitution model in Korean voice telephony markets. Freie Universität Berlin, *Academic research, 3*.

Armstrong, M. (2002). The theory of access pricing and interconnection, in: M. Cave, S. Majumdar, and I. Vogelsang (eds.), Handbook of Telecommunications Economics.

Baker, J.B., & Bresnahan, T.F. (1985). The gains from merger or collusion in product-differentiated industries. *The Journal of Industrial Economics, 33*(4), 427-444.

Baker, J.B., & Bresnahan, T.F. (1988). Estimating the residual demand curve facing a single firm. *International Journal of Industrial Organization, 6*(3), 283-300.

Bresnahan, T. F. (1982). The oligopoly solution concept is identified. *Economics Letters, 10(1)*, 87-92.

Bresnahan, T. F. (1989). Empirical studies of industries with market power. *Handbook of industrial organization, 2*, 1011-1057.

BEREC. (2012). Report on the impact of fixed-mobile substitution on market definition. BoR (12)52

Briglauer, W., Schwarz, A., & Zulehner, C. (2011). Is fixed-mobile substitution strong enough to de-regulate fixed voice telephony? Evidence from the Austrian markets. *Journal of Regulatory Economics, 39(1)*, 50-67

Briglauer, W., Götz, G., & Schwarz, A. (2010). Can a margin squeeze indicate the need for deregulation? The case of fixed network voice telephony markets. *Telecommunications Policy, 34(10)*, 551-561.

Calzada, J., & Valletti, T. M. (2008). Network Competition and Entry Deterrence. *The Economic Journal*, 118(531), 1223-1244.

Davis, P., & Garcés, E. (2009). Quantitative techniques for competition and antitrust analysis. Princeton University Press.

Ecorys. (2013). Future electronic communications markets subject to ex-ante regulation. Study for the European Commission. Retrieved from http://ec.europa.eu/information_society/newsroom/cf/dae/document.cfm?doc_id=3148

Enders, W. (1995). Applied Econometric time series: Wiley.

Finnish Communications Regulatory Authority/PÖYRY. (2009). Mobile phone service prices 2009 – international comparison. 3/2009.

Forchheimer, K. (1908). Theoretisches zum unvollständigen Monopole. *(Schmollers) Jahrbuch für Gesetzgebung, Verwaltung und Volkswirtschaft, 32*, 1–12

Genakos, C., & Valletti, T. (2011). Testing the "waterbed" effect in mobile telephony. *Journal of the European Economic Association, 9(6)*, 1114-1142.

Gans, J. S., & King, S. P. (2005). Competitive neutrality in access pricing. *Australian Economic Review, 38(2)*, 128-136.

Gassner, K. (1998). An estimation of UK telephone access demand using Pseudo-Panel data. *Utilities Policy, 7(3)*, 143-154.

Growitsch, C., Marcus, J., & Wernick, C. (2010). The effects of lower mobile termination rates (MTRs) on retail price and demand. *Communications and Strategies, (80)*, 119-140.

Granger, C.W., & Newbold, P. (1974). Spurious regressions in econometrics. *Journal of econometrics, 2*(2), 111-120.

Gyimah-Brempong, K., & Karikari, J. A. (2007). Telephone Demand and Economic Growth in Africa. OECD Economic Studies, 38(1).

Inderst, R., & Peitz, M. (2012). Network investment, access and competition. *Telecommunications Policy, 36(5)*, 407-418.

Johnston, J., & DiNardo, J. (1997). Econometric Methods 4th Edition McGraw-Hill: New York.

Hjalmarsson, E., & Österholm, P. (2007). Testing for cointegration using the Johansen methodology when variables are near-integrated: International Monetary Fund.

Kahai, S.K., Kaserman, D.L., & Mayo, J.W. (1996). Is the Dominant Firm Dominant-An Empirical Analysis of AT&T's Market Power. *Journal of Law and Economics, 39(2)*, 499-517.

Laffont, J. J., & Tirole, J. (2001). Competition in telecommunications. MIT press.

Lerner, A. P. (1934). The concept of monopoly and the measurement of monopoly power. *The Review of Economic Studies, 1(3)*, 157-175.

McCloughan, P., & Lyons, S. (2006). Accounting for ARPU: New evidence from international panel data. *Telecommunications Policy, 30(10)*, 521-532.

Nardotto, M., Valletti, T. M., & Verboven, F. (2012). Unbundling the incumbent: Evidence from UK broadband. Centre for Economic Policy Research.

Olsson, O. (2011). Estimating the demand and market power of a firm in sawn wood markets. *Aalto University Working Paper 1194*.

Qu, F. (2007). Empirical Assessment of Market Power in the Alberta Wholesale Electricity Market. *Energy Studies Review, 15(1)*, 4

Scheffman, D. T., & Spiller, P. T. (1987). Geographic market definition under the US Department of Justice Merger Guidelines. *Journal of Law and Economics, 30*.

Scherer, F.M., & Ross, D. (1990). Industrial market structure and economic performance. Houghton Mifflin.

Suslow, V. Y. (1986). Estimating monopoly behavior with competitive recycling: an application to Alcoa. *The RAND Journal of Economics*, 389-403.

Vickers, J. (2005). Abuse of Market Power. *The Economic Journal, 115(504)*, 244-261.

Vogelsang, I. (2003). Price regulation of access to telecommunications networks. *Journal of Economic Literature, 41(3)*, 830-862.

Vogelsang, I. (2010). The relationship between mobile and fixed-line communications: A survey. *Information Economics and Policy, 22(1)*, 4-17.

Ward, M. R., & Woroch, G. A. (2010). The effect of prices on fixed and mobile telephone penetration: Using price subsidies as natural experiments. *Information Economics and Policy, 22(1)*, 18-32.

Worcester, D.A. (1957). Why" dominant firms" decline. *The Journal of Political Economy, 65*(4), 338-346.

Yule, G.U. (1926). Why do we sometimes get nonsense-correlations between Time-Series? *Journal of the royal statistical society, 89*(1), 1-63.

9. Technical Annex

9.1. Estimation of reference model

In this technical annex, the detailed tests leading to the stable and interpretable model reported in the paper are presented. The Stata programme code to reproduce these results is reported in footnotes.

Second stage estimation (reference model)

For convenience the basic model equation (6) representing residual demand can be restated here.

$$q_D = \phi_0 + \phi_1 \hat{p} + w_F' \phi_F + w_C' \phi_C + x' \phi_B + \varepsilon$$

In this model no intertemporal effects are considered, i.e. changes of variables only affect the dependent variable of the same period. The second stage 2SLS estimations (robust standard errors) for residual demand using instrumented prices fitted in the first stage (see following section) are reported in Table 3. The estimation represents a regression of dominant firm quantities (q_D) on instrumented prices ($\hat{p}$) and other variables[41].

q_D	Variable	Coefficient estimate	Std. err.	P>|t|
$\hat{\phi}_1$	Price	-0.663	0.364	0.068
$\hat{\phi}_{B1}$	GDP per capita	-0.333	0.726	0.646
$\hat{\phi}_{B2}$	Number of active PSTN access lines	1.980	0.366	0.000
$\hat{\phi}_{B3}$	Seasonal dummies for quarters 2, 3 and 4	-0.032	0.016	0.042
$\hat{\phi}_{B4}$		-0.065	0.024	0.007
$\hat{\phi}_{B5}$		-0.021	0.013	0.112
$\hat{\phi}_{F1}$	Voice origination prices per minute	0.150	0.133	0.259
$\hat{\phi}_{F2}$	Number of wholesale ADSL lines	-0.220	0.043	0.000
$\hat{\phi}_{F3}$	Number of unbundled accesses	-0.001	0.001	0.548
$\hat{\phi}_{F3}$	Mobile termination rate (average)	0.116	0.071	0.104
$\hat{\phi}_{F4}$	Interest rates 30y Govenment bond	-0.050	0.052	0.338
$\hat{\phi}_{F5}$	Exchange rates EUR/CHF	-0.220	0.155	0.155
$\hat{\phi}_0$	_constant	4.390	4.102	0.285
R^2	0.9658			
Prob>F	0.000			
N	28			

Table 3 - Second stage estimation of residual demand (reference model)

---

[41] Stata command: ivregress 2sls lntradvoipallminscm lnyrealcapita lnpstn d2 d3 d4 lnscmregorr lnadslwhole lnullreal lnwavgmtrr lninterest30y lnfxratechfeur (lntradvoipallarpmallr=lnSCMstaff lnadslretail), first robust

Note that as preliminary regressions have indicated possible serial correlation of errors, robust standard errors are used (Huber-White-Sandwich estimator)[42].

First stage estimation (reference model)

Similarly, for convenience basic model equation (5) representing instrumented prices can be restated here.

$$\hat{p} = \theta_0 + w_D' \theta_D + w_F' \theta_F + w_C' \theta_C + x' \theta_B + \varepsilon$$

The first stage 2SLS estimations (robust) for the regression of price on the instruments in the system are reported in Table 4.

p	Instruments	Coefficient estimate	Robust Std. err.	P>|t|
$\hat{\theta}_{B1}$	GDP per capita	-1.450	0.620	0.035
$\hat{\theta}_{B2}$	Number of active PSTN access lines	2.063	1.280	0.129
$\hat{\theta}_{B3}$	Seasonal dummies for quarters 2, 3 and 4	0.033	0.020	0.133
$\hat{\theta}_{B4}$		0.052	0.019	0.015
$\hat{\theta}_{B5}$		0.005	0.033	0.884
$\hat{\theta}_{F1}$	Voice origination prices per minute	0.381	0.189	0.063
$\hat{\theta}_{F2}$	Number of wholesale ADSL lines	-0.363	0.231	0.139
$\hat{\theta}_{F3}$	Number of unbundled accesses	-0.003	0.003	0.353
$\hat{\theta}_{F4}$	Mobile termination rate (average)	0.111	0.167	0.517
$\hat{\theta}_{F5}$	Interest rates 30y Govenment bond	0.040	0.09	0.664
$\hat{\theta}_{F6}$	Exchange rates EUR/CHF	-0.261	0.257	0.327
$\hat{\theta}_{D1}$	Number of staff working for Swisscom	-0.225	0.185	0.243
$\hat{\theta}_{D2}$	Number of retail ADSL lines active for Swisscom	0.331	0.285	0.266
$\hat{\hat{\theta}}_0$	_constant	-21.6948	10.074	0.049
R^2	0.9298			
Prob>F	0.000			
N	28			

Table 4 - First stage estimation of residual demand (reference model)

What is of importance in this regression that there are significant instruments. This is the case.

9.2. Identification of econometric problems

Time series models are particularly complex to interpret as various time related problems can negatively affect the stability of the model and the extent to which it can be interpreted. Most importantly, two potential problems will need to be addressed: serial correlation of errors and non-stationarity of variables. This technical chapter analyses these problems and presents the solution adopted in the form of an ARDL model implying that the baseline equations are added a one period lagged dependent and independent variables.

---

[42] The estimations can be replicated (when omitting the robust command) by running a simple regression of the instruments on the instrumented variable, saving the fitted values and regressing the dependent variable of the second stage against the explanatory variables as well as the fitted variable. With robust standard error estimates though, results would slightly defer.

Similar papers to this have often not proposed extended econometric analysis to check for the stability of results. In what follows it will be tried to address these problems in a structured way. Econometric modelling often implies a necessity for compromise as what desired by theory cannot be estimated efficiently. It will be shown that even though the sample size is limited, stable and meaningful results can be obtained.

When analysing the time series of the strategic variables graphically, it appears that there is a sufficient degree of variation in the locations of p and q_D to explain a demand and supply relationship (Figure 9).

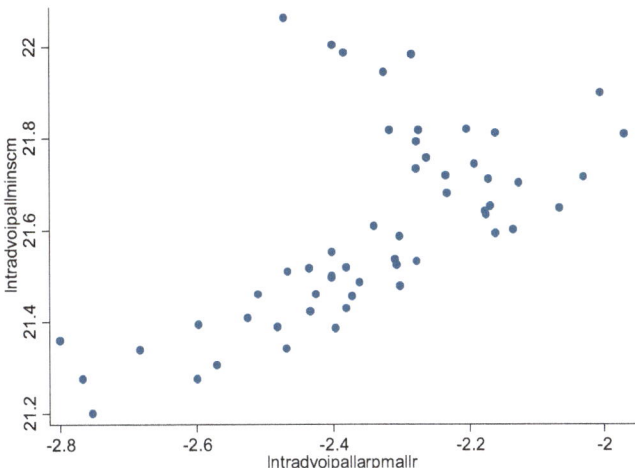

Figure 9 – Residual demand and prices in values[43]

The above graph shows the observed market outcomes (price is plotted on the horizontal axis and residual demand on the vertical). In the two following sections a general analysis of the time series under consideration is conducted and tests for the presence of non-stationarity and co-integration as well as serial correlation of errors are performed. Then, the econometric issues are analysed in detail and solution concepts are elaborated.

9.3.   Test for serial correlation of errors

In case of serial correlation of errors, the errors tend to depend on their own lagged values. In the following it will be verified whether this is the case in the present regressions. The serial correlation of the second stage residuals is reported in Figure 10.

---

[43] twoway (line lntradvoipallminscm quarternew)(line lntradvoipallminoth quarternew)

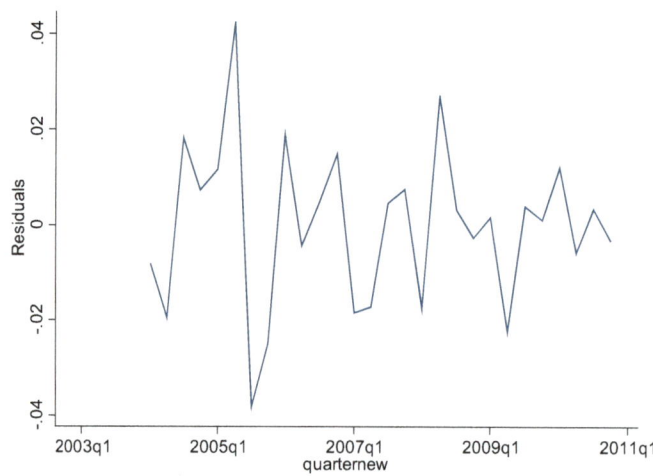

Figure 10 – Graph of residuals over the period under consideration

From a graphical analysis there could be serial correlation of residuals, but the effect does not seem to be strong. A Portmanteau test for white noise confirms serial correlation of errors in the first stage regression (Table 5).

Portmanteau test for white noise [44]
Portmanteau (Q) statistic =   25.1171
Prob > chi2(12)   =   0.0143

Table 5 – Portmanteu test for serial correlation (first stage equation)

The test suggests rejecting the null hypothesis of no autocorrelation of errors (at a 5% critical value). There seems therefore clearly to be a problem of serial correlation.

9.4.   Test for non-stationarity

It is important to know whether the series under consideration are stationary, i.e. have constant mean, variance and covariance over time or not, as this changes the framework of analysis required. The development over time of the dominant firm and fringe output is illustrated in Figure 11 (incumbent in blue, fringe in red), while the development in first difference is reported in Figure 12. The development of price levels of both Swisscom and competitor prices is represented in Figure 14 (incumbent in blue, fringe in red). Finally, Figure 13 reports the development of market shares of Swisscom.

---

[44] Stata command: wntestq res5 (where residuals are predicted after second stage using the true values of the endogenous variable.

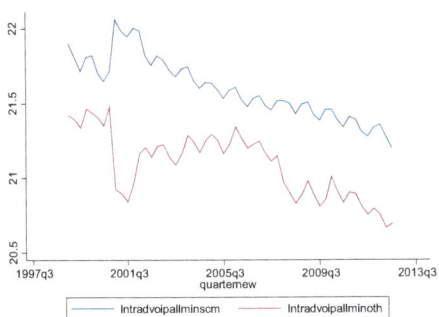

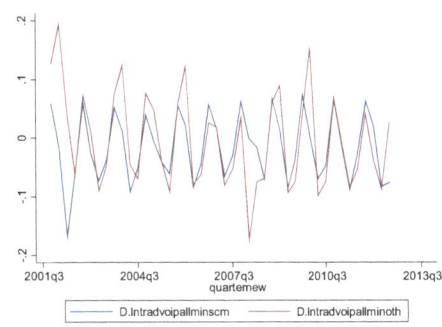

Figure 11 – Dominant firm and fringe outputs (national voice traffic minutes) [45]

Figure 12 - Dominant firm and fringe outputs (national voice traffic minutes), first differences

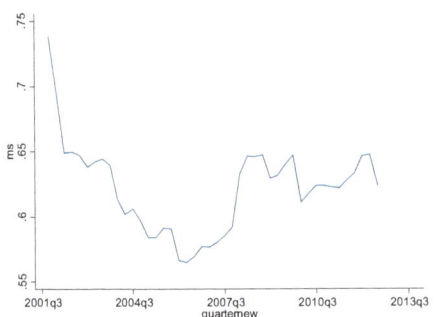

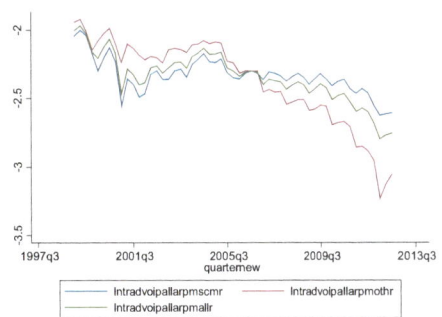

Figure 13 – Dominant firm market share on voice traffic minutes

Figure 14 – Average prices (revenues) per minute

Figure 11 shows a general decline of the fixed telephony market which concerns both the incumbent as well as competitors. Overall, the output series are unlikely to be stationary (mean is decreasing over time). When looking at first differences it can be seen that there seem to be no changes in mean or variance (of first differences) over time. Figure 13 reports Swisscom's market share in detail (here for fixed national outgoing traffic only) which is in a declining market relatively stable over time as shown in the introduction on aggregate data. Finally, the analysis of prices from Figure 14 seem to suggest a decreasing trend and that except for the few last quarters under observation it may be reasonable to consider a single market price as prices move relatively closely together and it can even be seen that from around mid-period (2006) the sign of price differences between operators change and Swisscom perceives slightly higher per minute prices. As for also quantity, prices seem to decline over time not suggesting a stable mean and therefore unlikely to be stationary.

When looking at correlograms to identify how strongly variables correlate with their own "past" (autocorrelation) it can be seen in the following figures that all important strategic variables are significantly correlated with their own first lag. There seems to be, therefore, some kind of feedback effect in the market which has not been taken into account in the baseline model[46].

---

[45] twoway (line d.lntradvoipallminscm quarternew)(line d.lntradvoipallminoth quarternew) if t>16
[46] AC=Autocorrelation, PAC=Partial autocorrelation

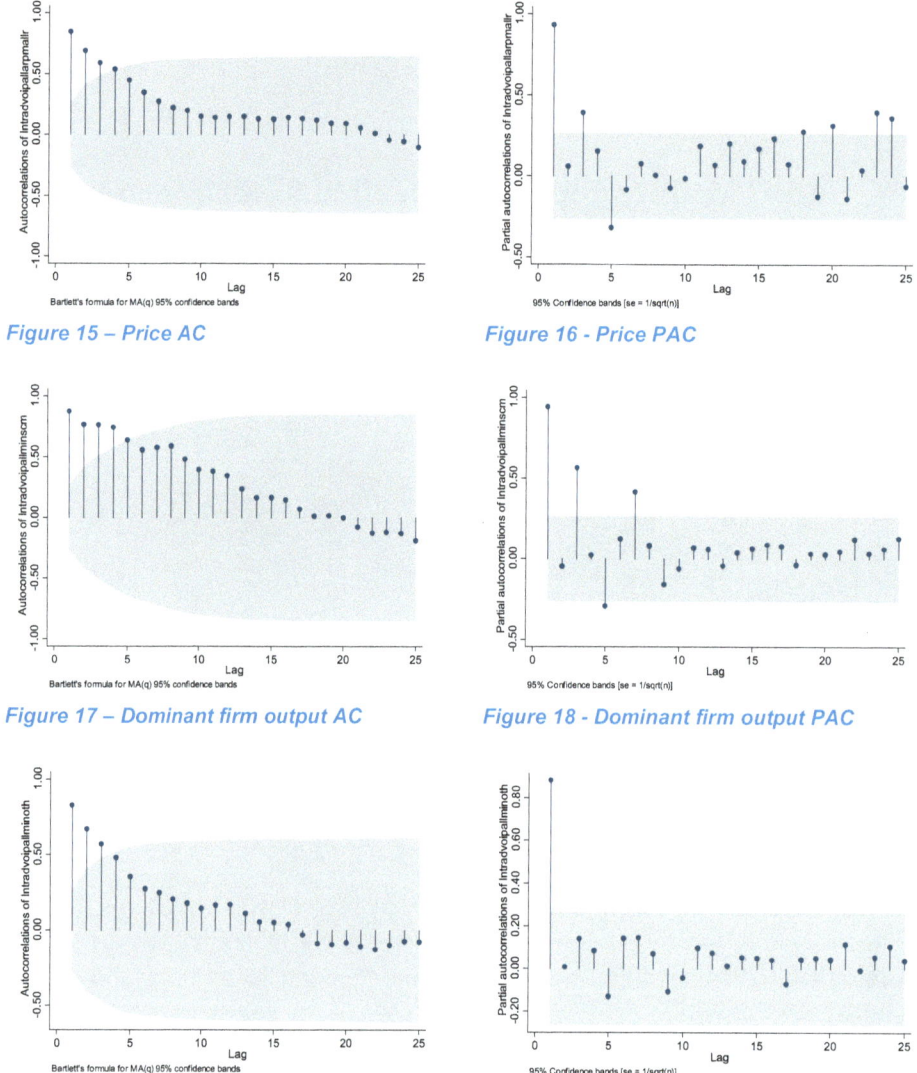

Figure 15 – Price AC

Figure 16 - Price PAC

Figure 17 – Dominant firm output AC

Figure 18 - Dominant firm output PAC

Figure 19 – Fringe output AC

Figure 20 – Fringe output PAC

While autocorrelations show that the correlation with current periods tends to decrease smoothly over time for all strategic variables considered, the partial autocorrelations (controlling for every single lag) show that the most important lags in explaining current levels is t-1, i.e. the preceding quarter for all variables analysed. Most importantly, Figure 17 and Figure 18 show that Swisscom's output strongly depends on its output in t-1 (it also depends relatively strongly on values in t-3). A similar effect applies to fringe output as can be seen in Figure 19 and Figure 20. This effect is also observed in Figure 15 and Figure 16 where prices seems to importantly depend on own values in the preceding period suggesting the same problem for the first stage regression. In general this analysis suggests that there is some kind of intertemporal effect which might need to be taken into account in the model.

Standard tests for stationarity include simple Dickey Fuller tests (no difference lags included). The Dickey Fuller test results are not reported here in detail. They suggest, however, that all variables used in the regression were, over the relevant time horizon of the regression, non- stationary, i.e. integrated of a higher order than zero. An exception is the variable "number of retail broadband subscribers of Swisscom" which is stationary as well as the constant term and the dummies). From all variables under examination in the period under review (4Q2003 to 3Q2012) all variables with an integration order higher than zero were, however, stationary in their differences and therefore integrated of order one. When performing an augmented Dickey Fuller test taking into account a series of lags these results are generally confirmed. The results overall indicate that there is a stationarity problem which has to be addressed.

9.5. <u>Addressing serial correlation of errors</u>

Under serial correlation of errors, the usual OLS estimators, although linear, unbiased and asymptotically normally distributed are no longer minimum variance among all linear unbiased estimators. As a result, the coefficient estimates remain valid but the usual t, F, and χ^2 statistics may be less efficient (higher probability of rejecting the true hypotheses and accepting wrong hypothesis). Serial correlation of errors may be caused among other things by a specification bias, excluding relevant variables from the regression (including lagged variables), non-stationarity of the series or incorrect functional form. There are different ways to address the problem of serial correlation. In the present model, the following approaches to structurally improve the model are of interest.

i) Using log-log specification
In some cases, the introduction of log-log specifications can improve results (as the procedure tends to shrink outliers) and lead to less serial correlation[47]. As logs are already used no further improvement is possible.

ii) Using first differences
The variables could all be first differenced. Such a transformation is, however, not compatible with the outlined model as coefficients of the level variables represent elasticities (e.g. residual demand elasticity). Differentiating these values would not allow the necessary inferences. For this reason this otherwise popular method is discarded.

iii) Adding new (lagged) variables
New variables, especially lagged variables could be introduced in order to effectively remove serial correlation of errors. Given the possibly dynamic nature of the model this could be a solution in one or both stage regressions[48]. Such transformations would be unlikely to change the model qualitatively and would only take into account the transmission effects of the function across time (in the baseline specification it is supposed that changes in t only have impacts in t). When analysing autocorrelation of the variables it is seen that dominant firm output seemed to be strongly correlated with its own first to third lag (the second lag is less correlated). Considering AC and PAC statistics the first lag in particular has been shown to be of particular importance.

Lagged variables (including the dependent variable) can be taken into account as exogenous (coming from t-1). Any introduction of a lagged dependent variable would mean that any other variable in the regression would have an "after effect" into the future. A change in any exogenous variable would then not only have a direct effect on the output demanded in t, but indirectly over the very change of the dependent variable in t also on, for example, the quantity demanded in t+1. A change in an independent variable would therefore result in an increase/decrease of the dependent variable to some extent also in t+1, t+2, etc. Long run multipliers measuring the long run marginal effects need in this case be calculated. Similarly it would also be possible to introduce lagged independent variables. This would then explicitly take into account the

---

[47] Baker and Bresnahan (1988)
[48] Olsson (2011)

dynamics of the effect of a specific shock on the dependent variable. Considering the above, using lagged dependent and independent variables would be broadly compatible with the baseline model, simply introducing intertemporal effects. It has, however, to be considered that the low number of observation constrains the possibilities to include a large number of additional variables.

Finally, the model may simply continue to be estimated by OLS as coefficients are still unbiased, but the standard error may be corrected to be again minimum variance (e.g. using Newey West or Huber-Sandwich-Sandwich corrections to make the estimation of the standard error minimum variance, as used in the preceding estimations). The standard error is then called "robust".

Overall, the option of introducing lagged variables seems a reasonable way to solve the problem. As such an option may, however, also have an influence on cointegration (see next section), possible adjustments to the model will be analysed subsequently.

9.6.   Addressing non-stationarity

As described in Granger and Newbold (1974) stationarity (integration of order zero) is one of the underlying assumptions of a linear regression. In case of non-stationarity, when for example two otherwise independent variable grow over time (i.e. the mean is not constant), a significant statistical relation between the two can be found in a standard OLS regression when none exists. This not because these variables would have something in common but simply because both are growing (and this growth may be driven by outside variables). Yule (1926) has shown that spurious correlation would even persist in non-stationary series which are long. This effect could be controlled when introducing trends into the regressions. However, the theoretical model should be able to explain growth over time rather than take it as exogenous and eliminate it. Introducing a trend would in the present case therefore not be appropriate.

If some variables in the regression are integrated of order one, then the usual statistical results may or may not be valid. Only in the case when regressors are integrated of order one (I(1)) and also "co-integrated", the regression can be reconstructed in a way that allows for valid inference without producing spurious results. Cointegration occurs when the long run stochastic trends in two processes are the same so they cancel. Variables then have a "common trend". In other words there must be a linear combination of the variables which is stationary (I(0)). Formally,

$$\beta_1 X_{1t} + \beta_2 X_{2t} + \cdots = 0$$

Where $\beta = (\beta_1, \beta_2, \ldots, \beta_n)$ is called the co-integrating vector. The equilibrium error is the difference in t to the common trend. The equilibrium error for a given co-integrating relationship is $e_t = \beta\ x_t$. If x_t has n components there may be as many as n-1 linearly independent co-integrating vectors. When x contains only two variables, only one co-integrating relationship is possible. A matrix B may then include all co-integrating vectors. The number of co-integrating vectors of x is then called the cointegration rank of x (Enders, 1995). Intuitively, in the case of two of more co-integrated variables a shock on one variable would indeed have a relevant long term effect on the other variable in which case a regression may again be valid due to a common stochastic trend. There may therefore be departures from the long run equilibrium trend, but such a trend exists and the values of the variables tend to it over time. If there is not at least one such common trend between I(1) variables regressions using non-stationary variables are spurious.

An Engle-Granger test can be performed to check for cointegration of variables. In particular, it checks whether the residuals of the regressions are stationary or not (regressing the first difference of the residuals in t on the lagged level of residuals in t-1)[49]. In case of stationarity, the variables are co-integrated. When performing the test (Table 6 and Table 7), the hypothesis of a unit root in the errors in the baseline model in the first as well as in the second stage cannot be rejected[50].

	Test Statistic	1% Critical Value	5% Critical Value	10% Critical Value
Z(t)	**-5.221**	-6.846	-5.945	**-5.512**

Critical values from MacKinnon (1990, 2010)

Table 6 - Engle-Granger test for cointegration for the first stage regression[51]

	Test Statistic	1% Critical Value	5% Critical Value	10% Critical Value
Z(t)	**-5.308**	-7.955	-6.857	**-6.340**

Critical values from MacKinnon (1990, 2010)

Table 7 - Engle-Granger test for cointegration for the second stage regression[52]

The errors are therefore non-stationary and it can be suspected that neither the first nor the second stage equations in the baseline model are co-integrated. The results suggest therefore that the baseline estimates are the result of a spurious regression and are not properly interpretable. When interpreting such spurious regressions (such as the reference estimates in this paper), it can be considered that the estimation bias is always in the direction of rejecting a true null hypothesis (Granger and Newbold (1974)). Therefore, inferences leading to accepting the null hypotheses should usually be correct. In the case of the t test on coefficients (H0 is that the coefficient equals zero) this would imply that a connection between variables could be found which are in reality non existing. Instead, decisions to not reject the hypothesis that the coefficient is 0 are likely to be correct.

Finally, from a technical point of view regressions are meaningful when sufficient cointegration can be supposed – this even when including formally not integrated (or near integrated) variables (Hjalmarsson & Österholm, 2007). The following sections show how the problem of non-stationarity and absence of cointegration can be solved.

9.7. <u>Addressing non-stationarity and lacking cointegration with an ARDL model</u>

---

[49] Methodologically this test differs only slightly from the Dickey Fuller tests performed on the variables to check for their stationarity (using McKinnon Criterion).

[50] Note that it is formally incorrect using the I(0) variable which is included in our model (number of ADSL retail lines of the incumbent) in this test (or even in the model). However, the test results differ only slightly when excluding the variable. At a 5% critical value the hypothesis of non stationarity of the errors (Z(t)= -5.326)) could not be rejected. Ideally regressions should only be run using I(1) variables which are co-integrated with each other (full rank according to Johansen). In practice this is, however, rarely the case and regressions are also run without full rank and with few I(0) variables.

[51] egranger lntradvoipallarpmallr lnadslwhole lnullreal lnwavgmtrr lninterest30y lnfxratechfeur lnSCMstaff lnadslretail

[52] egranger lntradvoipallminscm xboriginal lnyrealcapita lnpstn lnscmregorr lnadslwhole lnullreal lnwavgmtrr lninterest30y lnfxratechfeur

In most cases, non-stationary variables are difference stationary I(1), implying that the integration order of the differences is zero. This is the case also with the variables used in this paper. This automatically means that when using first differences instead of levels the non-stationarity problems of the regression disappear. Taking first differences is therefore a popular solution to the stationarity problem. As already described in the section on serial correlation of errors, the problem of this approach in the present paper is that information on levels are lost. While with a level estimation the estimated coefficients represent elasticities, in a first difference model coefficients would only indicate acceleration of such coefficients over time. Even in steady state, there would be no further information. Such an approach would, therefore, not be useful to determine residual demand elasticity and market power and is discarded.

This paper instead uses an Auto-Regressive Distributed Lags model to address the non-stationarity problem. Similarly to the autocorrelation issue the non-stationarity problem may be overcome by introducing lags of the dependent and independent variables, as long as there is cointegration. A simple first order ARDL model may represent a convenient bridge between a purely theoretical level model without lags and a more pragmatic data analysis considering intertemporal feedback effects. With dependent variable y and independent variables z, an ARDL (1,1) model, i.e. including one lag of the dependent variable and one lag of the independent variable, can be used (see Equation (9)).

$$y_t = c + \alpha_1 y_{t-1} + \beta_0 z_t + \beta_1 z_{t-1} + \epsilon_t \tag{9}$$

In (9) β_0 is called the "impact multiplier" (i.e. the immediate same period effect of the explanatory on the dependent variable). The other coefficients are dynamic multipliers representing how the system will adjust to the shock over time. The main objective of this paper is to identify the long term effects of changes in explanatory variables. The steady state is reached when the following equations hold.

$$y_t = y_{t-1} = y^*; z_t = z_{t-1} = z^* \text{ and } \epsilon_t = 0$$

Then, the steady state is represented by the following equation:

$$y^* = \frac{c}{1 - \alpha_1} + \frac{\beta_0 + \beta_1}{1 - \alpha_1} z^*$$

In an ARDL(1,1) model a permanent change in z therefore affects y over all future periods, increasing it permanently. This overall impact, the "long run multiplier" is represented by (10).

$$\frac{\partial y^*}{\partial z^*} = \frac{\beta_0 + \beta_1}{1 - \alpha_1} \tag{10}$$

Equation (10) can be estimated when the coefficients of the ARDL regression are known.

A possible problem when using ARDL for the model considered here could be that if the lagged endogenous variable appears on the right-hand side of the regression equation (as in (9)) and the disturbances continue to be autocorrelated, then the lagged endogenous variable will automatically be correlated with the disturbance term and thus become endogenous leading to biased and even inconsistent results. It the following sections it is, however, shown that the introduction of ARDL not only leads to sufficient cointegration but that in such a model errors are also not serially correlated anymore, meaning that this problem is inexistent.

Adaption of an ARDL Model

Given the nature of the theoretical model to be estimated (two equations, each with a series of explanatory variables to be estimated in two stages), it will first need to be analysed what type of ARDL model is most suitable. Then, it needs to be assessed whether introducing ARDL for both equations makes the first stage regression cointegrated and whether, therefore, a valid instrumented variable can be obtained. If this is the case, it also needs to be verified whether the second stage regression remains valid. Finally, it has to assessed whether there is still a problem of serial correlation of errors.

Choice of type of ARDL model

A first check of the relevant criteria to choose the number of lags to include for optimal cointegration (Table 8) suggests adding two to four time lags in the model (of the second stage equation using standard fitted values). This is not surprising as often in time series such tests suggest to introduce the maximum number of lags.

Selection-order criteria
Sample: 2005q1 - 2010q4   (N=24)

Lag	LL	LR	df	p	FPE	AIC	HQIC	SBIC
0	369.533				4.60E-26	-29.9611	-29.8309	-29.4702
1	609.76	480.45	100	0.000	8.10E-31	-41.6466	-40.2142	-36.2472
2	3850.12	6480.7	100	0.000	1.e-140*	-303.343	-300.609	-293.035
3	7335.47	6970.7	100	0.000		-591.289	-588.164	-579.508
4	7578.64	486.35*	100	0.000		-611.553*	-608.428*	-599.773*

Endogenous: lntradvoipallminscm xboriginal lnyrealcapita lnpstn lnscmregorr lnadslwhole lnullreal lnwavgmtrr lninterest30y lnfxratechfeur
Exogenous: _cons

Table 8 – Different criteria values to choose the number of lags to be considered[53]

Given the limited amount of data, the only *ARDL(k,k)* model that can be tested though is *k=1*, as including more lags in the model would make estimation unfeasible. Introducing only one time lag may not be optimal in resolving the problem of lacking cointegration, but there is a possibility that the problem is resolved. ARDL(1,1) regressions for the whole system are therefore run. In the following, it is shown that this model probably provides for sufficient cointegration.

Testing the first stage ARDL (1,1) regression for cointegration

The first stage equation, from (5) can be restated in ARDL(1,1) form in equation (11).

$$p_t = \hat{\theta}_0 + \hat{\theta}_1 p_{t-1} + w'_{D,t}\widehat{\theta_{D,t}} + w'_{D,t-1}\widehat{\theta_{D,t-1}} + w'_{F,t}\widehat{\theta_{F,t}} + w'_{F,t-1}\widehat{\theta_{F,t-1}} + w'_{C,t}\widehat{\theta_{C,t}} + \\ w'_{C,t-1}\widehat{\theta_{C,t-1}} + x_t'\widehat{\theta_{B,t}} + x_{t-1}'\widehat{\theta_{B,t-1}} + \epsilon_t \tag{11}$$

The Engle-Granger test used in the preceding section is unfortunately limited to testing regressions with a low number of variables included. A simple test for cointegration as in the preceding section is therefore not possible here[54]. The alternative Johansen test on the regression (Table 9) shows, however, that there are six cointegration relations in the first stage regression, implying a relatively large amount cointegration relationships (with respect to the eleven variables included) providing evidence that spurious results may be avoided. This is particularly true with small samples. The Johansen test indicates, that the cointegration

---

[53] varsoc lntradvoipallminscm xboriginal lnyrealcapita lnpstn lnscmregorr lnadslwhole lnullreal lnwavgmtrr lninterest30y lnfxratechfeur
[54] Some authors also use simple Dickey Fuller tests to check for stationarity of the residuals.

rank is relatively high. The results therefore show that a model taking into account one lag of all variables provides for sufficient cointegration and estimation of the instrumented variable should be valid[55].

Lags = 1
Trend: constant
Sample: 2004q2 - 2010q4, N =27

maximum rank	parms	LL	eigenvalue	trace statistic	critical value
0	11	532.1482		544.8985	277.71
1	32	608.7789	0.99657	391.6371	233.13
2	51	663.0072	0.98199	283.1806	192.89
3	68	699.7184	0.93408	209.7582	156
4	83	726.9439	0.86691	155.3071	124.24
5	96	752.8149	0.85286	103.5651	94.15
6	107	772.2581	0.76313	**64.6787***	68.52
7	116	783.7952	0.57454	41.6046	47.21
8	123	794.3609	0.54281	20.4732	29.68
9	128	799.8902	0.33607	9.4145	15.41
10	131	803.885	0.25614	1.425	3.76
11	132	804.5975	0.05141		

Table 9 – Johansen test for cointegration (first stage regression) [56]

Testing the second stage ARDL (1,1) regression for cointegration

The second stage equation from (6) can be restated in ARDL(1,1) form in (12):

$$q_{D,t} = \hat{\phi}_0 + \hat{\phi}_{1,t}\hat{p}_t + \hat{\phi}_{1,t-1}p_{t-1} + \hat{\phi}_{2,t-1}q_{t-1} + w'_{F,t}\hat{\phi}_{F,t} + w'_{F,t-1}\hat{\phi}_{F,t-1} + w'_{C,t}\hat{\phi}_{C,t} + w'_{C,t-1}\hat{\phi}_{C,t-1} + x_t'\hat{\phi}_{B,t} + x_{t-1}'\hat{\phi}_{B,t-1} + \epsilon_t \tag{12}$$

The Johansen test for this regression is reported in Table 9 and shows for the second stage that there are four co-integration relations (ten variables have been included).

Lags = 1
Trend: constant
Sample: 2004q2 - 2010q4, N =27

maximum rank	parms	LL	eigenvalue	trace statistic	critical value
0	10	482.4242		384.8823	233.13
1	29	544.5308	0.98995	260.6692	192.89
2	46	579.4391	0.92466	190.8526	156
3	61	609.6892	0.89362	130.3523	124.24
4	74	630.6383	0.78813	**88.4542***	94.15
5	85	646.6574	0.69474	56.4159	68.52
6	94	656.4064	0.51429	36.9178	47.21
7	101	663.9514	0.42815	21.8279	29.68
8	106	669.2342	0.32383	11.2624	15.41

[55] Although Johansen's methodology is typically used in a setting where all variables in the system are I(1), having few stationary variables in the system is theoretically not an important issue and Johansen (1995) states that there is little need to pre-test the variables in the system to establish their order of integration. If a single variable is I(0) instead of I(1), this will reveal itself through a co-integrating vector whose space is spanned by the only stationary variable in the model.
[56] Stata command: vecrank lnradvoipallarpmallr lnyrealcapita lnpstn lnscmregorr lnadslwhole lnullreal lnwavgmtrr lninterest30y lnfxratechfeur lnSCMstaff lnadslretail, lags(1)

9	109	673.9483	0.29475	1.8341	.	3.76
10	110	674.8654	0.06567			

Table 10 - Johansen test for cointegration (second stage regression) [57]

The results also show that in the second stage there is a consistent amount of co-integrating relationships and that the estimates of the second stage should also be valid.

Testing the ARDL (1,1) equations for serial correlation of errors

To conclude, it should be verified whether the ARDL (1,1) regressions continue to suffer the same problems of serial correlation of errors as the baseline model. All one period lagged variables (t-1) newly included in the regression can be considered exogenous, both in case the current period variables are exogenous or endogenous. The lagged dependent variable in the first stage (price) can therefore also be considered as an exogenous instrument ("coming" from the past)[58].

When testing the first stage instrumented variable equation for serial correlation of errors, no such serial correlation is found at critical values of 5% (Table 11).

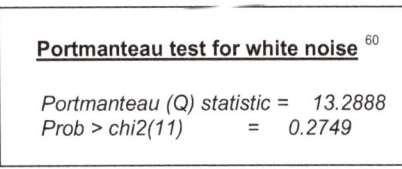

Portmanteau test for white noise [59]

Portmanteau (Q) statistic = 18.2777
Prob > chi2(11)   =   0.0754

Table 11 - Portmanteu test for serial correlation (first stage ARDL equation)

Overall, the first stage ARDL (1,1) regression is therefore estimating instrumental variables with instruments that are non-stationary, but having a sufficient degree of cointegration. In addition, there is no serial correlation of errors. The fitted value estimates produced in the first stage are therefore based on coefficients of a valid OLS regression, unbiased and efficient. It should, however, be noted that the significance of the instruments is reduced with respect to the first stage baseline estimates (given the now large number of variables this is not surprising).

As a final test for stability, a test for serial correlation of errors of the second stage regression is performed (see   Table 12).

Portmanteau test for white noise [60]

Portmanteau (Q) statistic = 13.2888
Prob > chi2(11)   =   0.2749

Table 12 - Portmanteu test for serial correlation (second stage ARDL equation)

---

[57] Staat command: vecrank lntradvoipallminscm lnyrealcapita lnpstn lnscmregorr lnadslwhole lnullreal lnwavgmtrr lninterest30y lnfxratechfeur xboriginal, lag(1)
[58] The lagged price variable is therefore represented by real and not fitted values.
[59] Stata command: wntestq ress (where residuals are predicted after first stage simple regression).
[60] Stata command: wntestq ress2 (where residuals are predicted after second stage using the true values of the endogenous variable).

The results indicate that not only in the first stage, but also in the second stage, serial correlation of errors is not an issue anymore when using the ARDL(1,1) specification. Nevertheless, standard errors continue to be estimated with a "robust" correction as before, as doing so may only further strengthen results.

The second stage ARDL (1,1) regression is therefore estimated with variables that are non-stationary, but having a sufficient degree of cointegration. In addition there is no serial correlation of errors.

Overall both the first and second stage ARDL (1,1) regressions are therefore valid. A full ARDL (1,1) model may consequently be used to correct the baseline model for the econometric problems identified.